Beyond Smoke and Mirrors
Climate Change and Energy in the 21st Century

What are the practical options for addressing global climate change?

How do we provide sustainable energy and electricity for a rapidly growing world population?

Which energy provision options are good, bad, and indifferent?

One of the most important issues facing humanity today is the prospect of global climate change, brought about primarily by our prolific energy use and heavy dependence on fossil fuels. Continuing on our present course using the present mix of fuels as the world economy and population grow will lead to very serious consequences. There are many claims and counterclaims about what to do to avert such potentially dire consequences. This has generated a fog of truths, half-truths, and exaggerations, and many people are understandably confused about these issues. The aim of this book is to help dispel the fog, and allow citizens to come to their own conclusions concerning the best options to avert dangerous climate change by switching to more sustainable energy provision.

The book begins with a composed and balanced discussion of the basics of climate change: what we know, how we know it, what the uncertainties are, and what causes it. There is no doubt that global warming is real; the question is how bad we will allow things to get. The main part of the book discusses how to reduce greenhouse gas emissions and limit the global temperature rise, including what the upper limit on greenhouse gases should be, how fast we should go to cut emissions, and all of the energy options being advocated to reduce those emissions. The many sensible, senseless, and self-serving proposals are assessed.

Beyond Smoke and Mirrors provides an accessible and concise overview of climate change science and current energy demand and supply patterns. It presents a balanced view of how our heavy reliance on fossil fuels can be changed over time so that we have a much more sustainable energy system going forward into the twenty-first century and beyond. The book is written in a non-technical style so that it is accessible to a wide range of readers without scientific backgrounds: students, policymakers, and the concerned citizen.

BURTON RICHTER is Paul Pigott Professor in the Physical Sciences Emeritus, and Director Emeritus, Stanford Linear Accelerator Center at Stanford University. He is a Nobel Prize-winning physicist for his pioneering work in the discovery of a heavy elementary particle. He received the Lawrence Medal from the US Department of Energy and the Abelson Prize from the American Association for the Advancement of Science. Over the past decade, he has turned his attention from high-energy physics to climate change and energy issues, and has earned a strong reputation in this field as well. He has served on many national and international review committees, but his most direct involvement is with nuclear energy where he chairs an advisory committee to the US Department of Energy. He is also a chairman of a recent American Physical Society study on energy efficiency, and a member of the "Blue Ribbon Panel" that oversaw the final edit of the US climate impact assessment that was released in 2000. He has written over 300 papers in scientific journals and op-ed articles for the *New York Times*, *Washington Post*, and *LA Times*.

Beyond Smoke and Mirrors

Climate Change and Energy in the 21st Century

BURTON RICHTER
Stanford University

CAMBRIDGE UNIVERSITY PRESS
Cambridge, New York, Melbourne, Madrid, Cape Town, Singapore,
São Paulo, Delhi, Dubai, Tokyo

Cambridge University Press
The Edinburgh Building, Cambridge CB2 8RU, UK

Published in the United States of America
by Cambridge University Press, New York

www.cambridge.org
Information on this title: www.cambridge.org/9780521763844

First published 2010

Printed in the United States of America

A catalog record for this publication is available from the British Library

Library of Congress Cataloging in Publication data
Richter, Burton, 1931–
    Beyond smoke and mirrors : climate change and energy in the 21st
    century / Burton Richter.
        p. cm.
    Includes bibliographical references and index.
    ISBN 978-0-521-76384-4 (hardback)
    1. Climatic changes–Forecasting.   2. Renewable energy sources–
    Forecasting.   3. Twenty-first century–Forecasts.   I. Title.
    QC903.R53 2010
    363.738′74–dc22   2010003023

ISBN 978-0-521-76384-4 Hardback
ISBN 978-0-521-74781-3 Paperback

Contents

Preface

This book is aimed at the general public and has been percolating in my head since mid-2006. It is not intended to be a textbook, but rather an accessible overview of what we know and don't know about climate change, what options we have to reduce greenhouse gas emissions in the energy sector of our economy, and what policies we should and should not adopt to make progress.

I am a latecomer to the climate and energy field. My career has been in physics. I received my PhD in 1956 and my Nobel Prize in 1976 at the relatively young age of 45. Many Nobel Laureates continue research, but some look for other mountains to climb, and I was one of those. I took on the job of directing a large Department of Energy scientific laboratory at Stanford University in 1984; its mission is to build and operate unique, large-scale research tools for the national scientific community. During my 15 years as director we expanded opportunities in many areas; the number of users from outside Stanford that came to the laboratory rose from about 1000 to nearly 3000, and the facilities that we pioneered were reproduced in many parts of the world.

Like many scientists, I had followed the growing debate on climate change from a distance, though I did have some peripheral involvement in related areas having to do with energy options. I became seriously interested in climate and energy issues in the mid-1990s, partly because it was clear that this would be a critical issue for the future and partly because of the lure of another mountain range. Since stepping down as a laboratory director in 1999, I have devoted most of my time to various aspects of the issue.

Having a Nobel Prize is a great advantage when moving into a new area. Besides being one of the highest scientific honors, it is a great door opener. Nobel Laureate Richter had a much easier time getting appointments with high-level officials in government and industry in the United States and abroad than would scientist Richter. I have served on many review committees, both national and international, ranging from the US government's analysis of the effects of climate change on the economy, to the nuclear energy programs of both the United States and France, to the role of efficiency in the reduction of greenhouse gas emissions.

The original 2006 outline for this book devoted much space to the reality of global warming. The pendulum has swung since then and the general public now seems convinced of its reality. Much credit for the change goes to former Vice President Al Gore, and to his movie and book *An Inconvenient Truth*. His Academy Award and Nobel Peace Prize are testaments to the influence of his work. His dramas have been important in getting people to pay attention, but for appropriate decisions to be taken, we need a more realistic view than his about the dangers, the uncertainties, and the opportunities for action.

The public needs and deserves an honest science-based explanation of what we know, how we know it, what the uncertainties are, how long it will take to reduce those uncertainties, and what we can do to reduce the risk of long-term changes to the world climate that make the Earth less hospitable to society. If I do my job well, the reader will have enough information to come to his or her own conclusion.

Personally, I should tell you that I do believe in beginning to invest in reducing greenhouse emissions as a kind of environmental insurance for my two young granddaughters (ages 5 and 2.5). A beginning now will cost much less than we are spending on the bailout of the world's financial institutions. If later information says that things are better or worse than we now expect, we can change our program, but the earlier we start the easier it will be to do some good.

Units

The book uses a combination of American and metric units. Almost all data on greenhouse gas emissions are given in metric units. Most electric power units are metric also. In this list I give some of the conversion factors.

TEMPERATURE

1 degree centigrade (C) = 1.8 degrees Fahrenheit (F)

LARGE NUMBERS

kilo (k) = thousand
mega (M) = million
giga (G) = billion (US) or thousand million (Europe)
tera (T) = thousand billion or a million-million
Examples: kilowatt (kW), gigatonnes (Gt), etc.

WEIGHT

tonne (t) = 1000 kilograms (kg) = 2200 pounds (lb)
ton = 2000 pounds

DISTANCE

1 meter = 39.4 inches
1 kilometer = 1000 meters = 0.62 miles

VOLUME

1 barrel (bbl) = 42 gallons (US)
1 liter = 1.056 quarts = 0.264 gallon

POWER

1 watt = basic unit of electrical power = 1 joule per second
1 gigawatt (GW) = one billion (or 1000 million) watts

ENERGY

Energy = power × time
1 kWh = 1 kilowatt-hour = 3 600 000 joules
1 BTU = 1054 joules
1 Quad = 1×10^{15} BTU = 1.054×10^{18} joules
1 TJ = 1×10^{12} joules

Conversion factors

Energy conversion factors

To:	TJ	Mtoe	MBTU	GWh
From:				
TJ	1	2.388×10^{-5}	947.8	0.2778
Mtoe*	4.1868×10^{4}	1	3.968×10^{7}	11 630
MBTU	1.0551×10^{-3}	2.52×10^{-8}	1	2.931×10^{-4}
GWh	3.6	8.6×10^{-5}	3412	1

Multiply *from* by *to* for number of units
*Million tonnes of oil equivalent

Mass conversion factors

To:	kg	t	ton	lb
From:				
kilogram (kg)	1	0.001	1.102×10^{-3}	2.2
tonne (t)	1000	1	1.1023	2204.6
ton	907.2	0.9072	1	2000.0
pound (lb)	0.454	4.54×10^{-4}	5.0×10^{-4}	1

Multiply *from* by *to* for number of units

Volume conversion factors

To: From:	gal US	gal UK	bbl	ft³	l	m³
US gallon (gal)	1	0.8327	0.02381	0.1337	3.785	0.0038
UK gallon (gal)	1.201	1	0.02859	0.1605	4.546	0.0015
barrel (bbl)	42.0	34.97	1	5.615	159.0	0.159
cubic foot (ft³)	7.48	6.229	0.1781	1	28.3	0.0283
liter (l)	0.2642	0.220	0.0063	0.0353	1	0.001
cubic meter (m³)	264.2	220.0	6.289	35.3147	1000.0	1

Multiply *from* by *to* for number of units

Abbreviations

ACEEE	American Council for an Energy Efficient Economy
AOGCM	atmosphere–ocean general circulation model
APS	American Physical Society
BAU	business as usual
BEV	battery-powered electric vehicle
CAFE	corporate average fuel economy
CCS	carbon capture and storage (sometimes sequestration)
CO_2	carbon dioxide, the main man-made greenhouse gas
CO_2e	carbon dioxide equivalent
DOE	US Department of Energy
DSM	demand side management
E_i	energy intensity (energy divided by GDP)
EGS	enhanced geothermal systems
EIA	Energy Information Administration (a division of the DOE)
EPA	US Environmental Protection Agency
EU	European Union
FF	fission fragments
GDP	gross domestic product
GNEP	Global Nuclear Energy Partnership
GRS	greenhouse gas reduction standard
HEU	highly enriched uranium (suitable for weapons)
IAEA	International Atomic Energy Agency
ICE	internal combustion engine
ICSU	International Council for Science
IEA	International Energy Agency (division of the OECD)
IIASA	International Institute of Applied Systems Analysis
IPCC	Intergovernmental Panel on Climate Change
LWR	light water reactor

NAS	National Academy of Sciences
NPT	Treaty on the Non-Proliferation of Nuclear Weapons
NRC	US Nuclear Regulatory Commission
OECD	Organization for Economic Co-operation and Development
OPEC	Organization of Petroleum Exporting Countries
OTA	Office of Technology Assessment
PHEV	plug-in hybrid electric vehicle
PPP	purchasing power parity
PV	photovoltaic
R&D	research and development
RPS	renewable portfolio standard
TCM	trillion cubic meters
TMI	Three Mile Island
TPES	total primary energy supply
TRU	transuranic elements
UK	United Kingdom
UN	United Nations
UNFCCC	United Nations Framework Convention on Climate Change
US	United States
VMT	vehicle miles traveled
WEC	World Economic Council
WMO	World Meteorological Organization
ZNE	zero net energy

1

Introduction

Our planet's atmosphere has been the dumping ground for all sorts of gases for as long as human history. When those using it as a dump were few, its capacity was large, and there was no problem. There are now more than six billion of us, and we have now reached the point where human activities have overloaded the atmospheric dump and the climate has begun to change. Our collective decision is what to do about it. Do we do nothing and leave the problem to our grandchildren who will suffer the consequences of our inaction, or do we begin to deal with it? It is much easier to do things now rather than later, but it will cost us something.

To me the answer is clear: we should start to deal with it. This book describes the problem and the alternatives that exist to make a start on limiting the damage. This is not an academic book, even though I am a physics professor. It is written for the general public. True, it does contain some scientific details for those interested in them, but they are in technical notes at the ends of chapters; you can skip them if you like.

The title of the book, *Beyond Smoke and Mirrors*, can be taken two ways. One is what future energy sources might replace coal and today's versions of solar power. The other is the real story behind the collection of sensible, senseless, and self-serving arguments that are being pushed by scientists, environmentalists, corporate executives, politicians, and world leaders. I mean the title both ways, and the book looks at the technical and policy options and what is really hiding behind the obscuring rhetorical smoke and mirrors. There are many ways to proceed and, unfortunately, there are more senseless arguments than sensible ones, and still more that are self-serving.

I divide those doing the most talking into the anti-greens, sometimes called the deniers; the greens; and the ultra-greens, sometimes called the exaggerators. As you might guess, I consider the greens to be the good guys. I classify myself among them.

There is a rapidly declining number of those denying that human activities are increasing the global temperature, but the species is not yet extinct and perhaps will never be. These are the anti-greens. Even they agree that the greenhouse effect is real, and that greenhouse gases in the atmosphere are the main element that controls the average temperature of the planet. Why they do not agree that changing the greenhouse gas concentration changes the temperature is beyond me.

The ultra-greens have declared an immediate planet-wide emergency where money is no object and where only solutions that match their programs are acceptable. They seem to have forgotten that the object is to cut greenhouse gas emission, not just to run the world on windmills and solar cells, which alone are insufficient to deal with the problem. By rejecting options that do not match their prejudices they make the problem more difficult and more expensive to address.

According to the anthropologists our first humanoid ancestor appeared about four million years ago. During the very long time from then until now the world has been both hotter and colder; the Arctic oceans have been ice-free before, and at other times ice has covered large parts of the world. What makes climate change a major problem today is the speed of the changes combined with the fact that there will be about nine billion of us by the middle of this century. We were able to adapt to change in the past as the climate moved back and forth from hot to cold, but there were tens of thousands of years to each swing compared with only hundreds of years for the earth to heat up this time. The slow pace of change gave the relatively small population back then time to move, and that is just what it did during the many temperature swings of the past, including the ice ages. The population now is too big to move *en masse*, so we had better do our best to limit the damage that we are causing.

Though there is now world agreement that there is a problem, there is no agreement on how to deal with it or even on what we should be trying to achieve. The European Union (EU), a collection of the richer countries, has a big program aimed at cutting greenhouse gas emissions. The richest country, the United States, has only recently acknowledged that human activity is the main cause of global warming, but has done very little so far to do anything about it. Russia thinks warming is good for it and has done nothing. The developing countries have said it is the rich countries that caused the problem so they should fix it, and poor countries should not be asked to slow their economic development. However, they are growing so fast that according to projections, the developing world will add as much greenhouse gas to the atmosphere in this century as the industrialized nations will have contributed in the 300 years from 1800 to 2100. We all live on the same globe, the actions of one affect all, and this problem cannot be solved without all working together.

There are three parts to this book. Part I is on climate change itself and explains what we know, what we don't know, what the uncertainties are in predictions of the future, and how urgent is the need for action. The section discusses what can be learned from the past, how the future is predicted, the many models that are used, and what they predict. The models are not yet good enough to converge on a single number for the expected temperature increase because the science is not that perfect. Uncertainty is used by some as an excuse for inaction, but it should not be, because by continuing "business as usual" the predictions for the end of the century range from terrible at the high end of the predicted increase (about 12 °F or 6 °C) to merely very bad at the low end (about 4 °F or 2 °C).

Part II begins with what we need to do in controlling greenhouse gas emissions to limit the ultimate temperature rise. It is too late in this century to return the atmosphere to what it was like before the start of the industrial age. I include my estimate of the allowable upper limit on greenhouse gases, the amount beyond which the risk of sudden climate instability greatly increases.

Next is a review of what the economists say about the best way financially of controlling emissions. There are no economists that I know who are saying do nothing now. The argument is over how fast to go. The natural removal time for the major greenhouse gases is measured in centuries, so if we wait until things get bad we will have to live with the consequences for a long time, no matter how hard we try to fix things. The issue is the problem that we will leave to our grandchildren.

Part II goes on to look at the sources of anthropogenic (human-caused) greenhouse gas emissions and what we might do about them. Two broad categories dominate: the energy we use to power our civilization; and agriculture and land use changes that have accompanied the increase in world population. I focus on energy use, which is responsible for 70% of greenhouse gas emissions. Agriculture and land-use changes contribute the other 30% of emissions, but their coupling to food production and the economies of the poor countries are not well understood. I leave this to others, except for biofuels which are part of the energy system.

I review what kinds of energy we use in the world economy and what each contributes to greenhouse gas emissions. The conclusion is the obvious one: fossil fuels are the culprit, and the only way to reduce their use while economic growth continues is by some combination of increased efficiency and a switch to sources of energy that do not emit greenhouse gases either by their nature or by our technology. In truth, we can continue our old ways of using fossil fuels for about another 50 years if we don't care about our grandchildren. Even with business as usual there are unlikely to be supply problems until the second half of the century, though there may be price problems.

There is no single technology that will solve all of our problems. We will have to proceed on many fronts simultaneously, starting with what we have in our technology arsenal now. All the options are reviewed, including capturing and storing away emissions from fossil fuels; efficiency; nuclear power; and all of those energy systems called the Renewables. Some are ready for the big time now, others need further development. All revolutionary technologies start in the laboratory, and we are also not investing enough in the development of the technologies of the future.

Energy supply is the area where one finds most of the senseless and self-serving calls to action. For example, it is not within the bounds of reality to eliminate all the fossil fuels from our electricity supply in the next 10 years. This one is senseless. Further, increasing the amount of corn-based ethanol in our gasoline does almost nothing to decrease emissions when emissions in ethanol production are included. This one is self-serving.

Part II concludes with an admittedly opinionated summary of the promise and the problems of various technologies (there are lots of both), as well as my personal scorecard showing winners, losers, and options for which the verdict is not yet in.

Part III concerns policy options. There are two dimensions that need discussion: what to do on a national or regional scale, and what to do on a world scale. I believe the best policies in market economies are those that allow the private sector to make the most profits by doing the right things rather than the wrong things. There is always a huge amount of brain power devoted to making money and it can and should be tapped. I call this "tilting the playing field" so that things move in a desired direction. Of course, regulations are required too. The US auto industry, for example, has resisted efficiency improvements until regulations required them to act. I know this area well, having spent six years in the 1980s on the General Motors Science and Technology Advisory Board. I think the industry has finally understood what is needed and I hope it survives the current economic downturn.

The global problem is harder to deal with. It is particularly tough because while emissions have to be tackled on a global basis, the world has countries that range from rich to poor. Most emissions are coupled to energy use, and energy use is coupled to economic development: the poor want to get rich, the rich want to get richer, and the benefits coming from actions now are going to be seen only in the future. The very poorest use so little energy that even as they begin to climb the development ladder and use more, they will still make only a tiny contribution to emissions, and the world program can leave them alone until they have climbed several steps.

But the developing countries in the rapid-growth phase – China and India, for instance – cannot be entirely left out of the action agenda as they were in the Kyoto Protocol of 1997. China has already passed the United States as the largest emitter of greenhouse gases, and the developing nations collectively are expected to surpass the industrialized ones in 15 to 20 years. There can be no effective program for the long term without all nations coming under a greenhouse control umbrella once they reach some emission threshold. It will no longer do for the developing nations to ask the industrialized nations alone to fix the problem, because they can't. In business-as-usual projections (continuing with the same mix of fuels as the world economy grows), the developing nations as a whole will emit nearly as much greenhouse gas from 2000 to 2100 as the industrialized nations will have done in the three centuries between 1800 and 2100. There is no solution to the global warming problem without the participation of the developing world. Policies have to reflect reality, and the richer counties will have to take the lead. There is no excuse for the United States to stand aside as it has done since 1997. The first Kyoto Protocol expires in 2012, and the new one now being worked on had better include some graduated way to include all but the very poorest nations.

In 1968 Garrett Hardin, then a professor of ecology at the University of California, Santa Barbara, published an enormously influential article, "The Tragedy of the Commons" [1]. The metaphor of the title referred to how overgrazing occurred on common pasture land in medieval England. It did no good for only one person to limit his sheep grazing because his contribution was so small. Only if all worked together to limit grazing could the common pasture be pre-served. Hardin's "Commons" today is the Earth's atmosphere.

We can preserve our atmospheric commons. What we know, how we know it, what the uncertainties are, and what we should be doing are the subjects of this book.

Part I Climate

2

Greenhouse Earth

If this were a science fiction story, I would tell of the underground cities on Mars that get their heat from the still-warm core of the planet. I would tell of the underground cities on Venus too, and their struggle to insulate themselves from the killing heat of the surface. In the story I would sympathize with the Martians because through no fault of their own, they lived on a planet that was too small to keep its atmosphere and with it the greenhouse effect that kept it warm enough for liquid water to flow. As for the Venusians, I would write with sadness of their blindness to the dangers of global warming and the runaway greenhouse effect that forced them underground.

There is no question about the reality of the greenhouse effect, even from those who still deny that human activities have anything to do with global warming. This chapter and the next three tell the story of how we can be so sure there is an effect, why almost everyone has finally concluded that our planet is getting warmer and that we are primarily responsible for it, and what the future holds if we continue on our present course.

To understand the issues, we can call on information about other planets in our Solar System as well as on what we can measure on our own. We and our nearest neighbors, Venus, closer to the Sun, and Mars, further out, are very much alike in composition, but very different in surface temperatures because of the greenhouse effect. Venus is too hot to support life, and Mars is too cold. Yet all were formed about 4.5 billion years ago, and all are made from the same stuff. The difference lies in their greenhouse effects: too much for Venus, too little for Mars, and just right for us.

The climate greenhouse effect is different in detail, but not in principle, from that which allows tomatoes to be grown in winter under a transparent roof. In the plant greenhouse, the transparent double-paned roof lets in sunlight and traps the heat that would otherwise escape. In the atmospheric greenhouse, greenhouse gases trap heat that would otherwise be radiated out into space. This is not complicated in principle, though it is complicated to calculate the surface temperature in the real world with precision (something that will be discussed later). Human activity that changes the greenhouse effect and traps more heat drives the concern about global warming. Even among the anti-green lobby there is no argument about the reality of the greenhouse effect, only about how human activity is changing it.

Our planet's average temperature is determined by a balance that is struck between the energy coming from the Sun and the energy radiated back out into space. What comes in depends on the temperature of the Sun, and what goes out depends on the Earth's surface temperature and on what things in the atmosphere block parts of the radiation. Think of it this way – what comes in from the Sun is almost all in the form of ordinary visible light. What goes out is mostly in the form of infrared radiation which we can't see but can certainly feel. If you have ever stood in front of an old-fashioned hot stove, you can feel the radiation coming from it though you cannot see it. This radiation is what is partially blocked by greenhouse gases and the temperature has to go up to let enough heat out through that part of the radiation window that remains open to balance what comes in from the Sun.

Carbon dioxide (CO_2) is the gas most discussed. It is the main man-made (anthropogenic) contributor, but it is not the only one (more about the others in Chapter 3). What is a surprise to most people is that none of the man-made gases contributes as much to keeping our planet warm as ordinary water vapor. (See Technical Note 2.1 if you are interested in more of the science of the greenhouse effect.)

Although the Earth has a core of molten iron, in our planetary greenhouse over 99.99% of the energy reaching the surface of our planet is sunlight. Rock is a very good insulator and a relatively small amount of heat from the interior reaches the surface. Glowing rivers of molten rock do come from volcanoes, but they cover a tiny fraction of the surface of our world and so contribute very little to the surface heat. What comes in is sunlight; what leaves is radiated heat called infrared radiation. We can ignore all the rest.

The total power incoming from the Sun dwarfs everything made by humans. The energy that comes in on the sunlit side of the Earth in one hour equals the total of all forms of energy used by mankind in one year. Sun power totals about 100 million gigawatts (1 GW equals 1 billion watts), equivalent to the energy output of 100 million large electricity generating plants. All the electrical power used in the

United States by everyone for everything totals only about 500 GW at the daytime peak in usage on a hot summer day.

It is simple to calculate roughly our planet's surface temperature if there were no greenhouse effect at all, though it is complicated to calculate what happens in the real world. With no greenhouse effect, none of the energy radiated would be blocked and only the surface temperature would determine the energy outflow needed to balance the energy from the Sun. If the system became out of balance, the temperature would change to bring it back. Too much heat leaving would cool the surface; too little would allow it to heat.

Assuming the entire surface of the Earth is the same, ignoring the difference between the day and night sides, ignoring the cold poles compared with the rest and assuming that nothing blocks the outgoing heat, the average temperature required to radiate enough to balance the incoming solar energy is –4 °F (–20 °C). A fancier calculation taking into account the things ignored in this simple calculation, but continuing with the assumption that nothing in the atmosphere blocks any of the radiated heat, gives a number only a few degrees higher.

The average temperature of the Earth is actually +60 °F (+15 °C). The difference of about 65 °F is entirely caused by the greenhouse effect, which traps part of the energy that would be radiated from the surface in its absence. The surface temperature has to increase so that the part of the radiated energy that can get through will carry enough energy to keep the system in balance. Without the greenhouse effect the Earth would be a frozen ball of slush. With it we have, on average, a comfortable world, capable of supporting diverse life forms.

Over the history of the Earth, the average temperature has varied considerably as the amount of greenhouse gases in the atmosphere has changed and as the output of the Sun has changed. Today, the concern about global warming focuses on human activity that causes an increase in some greenhouse gases. The logic is simple: greenhouse gases are known to increase the temperature, and if we add more of what increases the temperature, we will increase the temperature more. How much more is the question that thousands of scientists are trying to answer.

Looking again at our two nearest neighbors in the Solar System, Venus and Mars, tells what happens when the greenhouse effect goes very wrong. I began this chapter with what a science fiction story might be like. Here is the real story. Both planets have been extensively studied from Earth by telescopes and radar, observed by orbiting spacecraft, and sampled by probes sent into the atmosphere (Venus) and by landers (Mars).

Venus is closest to us in size. Its diameter is 94% of ours and its surface gravity is 92% of ours. If you weigh 150 pounds (68 kg) here, a scale would show that you weigh only 138 pounds (63 kg) there. However, you would not enjoy a trip to the Venusian spa. Venus is closer to the Sun, having an orbit that is 72% the size of ours (Earth's orbit radius is about 93 million miles or 150 million kilometers, that of Venus is about 70 million miles or 112 million kilometers, and that of Mars is about 140 million miles or 225 million kilometers). This means that the incoming radiation on Venus is almost twice as intense as on Earth and has to be balanced by a higher temperature to radiate enough energy to maintain a constant temperature. In the absence of any greenhouse effect, the average temperature of Venus would be about 90°F (32°C), uncomfortable, but livable at least in the cooler areas near the poles. Instead, its temperature is above 800 °F (450 °C), way above the temperature reached in a self-cleaning oven. Venus has what is called a runaway greenhouse effect. Though the cause of the runaway on Venus is not fully understood, we do know the consequences. Today the Venusian atmosphere is about 90 times denser than ours and it consists almost entirely of CO_2. The greenhouse effect is huge and so is the surface temperature.

Mars is smaller than Earth. Its diameter is 53% of ours and its surface gravity is 40% of ours. The person weighing 150 pounds on Earth, and 138 pounds on Venus, would weigh only 60 pounds (27 kg) on Mars. This low gravity is what makes Mars lifeless today. Mars has an orbit that is one and a half times as large as ours and so receives much less sunlight. Its surface temperature in the absence of any greenhouse effect would be about 65°F below zero (−54 °C). However, from the data transmitted from the Martian Rovers that traveled its surface in 2008 and 2009, we think that the red planet once had liquid water. It could only have had that if a greenhouse effect had once kept the temperature above the freezing point of water, +32 °F (0 °C). Unfortunately for Mars, its low gravity let its atmosphere (now only 1% of the density of Earth's atmosphere) diffuse away into space and with it went its greenhouse effect.

My science-fiction Martians knew their fate and mourned it; my Venusians ignored their fate and regret it, but this is science, not sci-fi, and the moral to be drawn about our activities and the greenhouse effect is that if you do not understand what you are doing, changing things on a global scale can be dangerous. Just how dangerous is the subject of Chapter 5. For now, remember that the Earth has been hotter and the Earth has been colder than it is today. Human existence has spanned many ice ages. The transition from Neanderthal man to *Homo sapiens* (us) as the dominant subspecies happened during the last ice age. But now there are 6 billion of us and during this century we will grow to 9 billion. The small human population of the last ice age could move around and go to where climatic conditions were tolerable. Moving 9 billion people is not going to be feasible. It is not our existence that is threatened by global warming, it is our civilization.

Technical Note 2.1: The science of the greenhouse effect

The energy radiated by an object is called "black-body" radiation. The name black-body was coined in the mid 1800s to describe objects that absorb all radiation that falls on them and reflect nothing. This can be confusing because the same term has come to describe two things: the black absorber, and the heat radiation from hot bodies like the Sun, for instance, that are anything but black.

If an object remains at a uniform temperature, the total amount of energy radiated away depends only on its size and its temperature. The wavelength at the peak of the radiation distribution also depends only on the temperature. Our Sun has a surface temperature of about 5800°kelvin (over 10 000°F) and a diameter of about 900 000 miles (about 100 times that of the Earth), and radiates a huge amount of energy. The peak in its radiation spectrum is at a wavelength of 0.5 micron (a micron is one-millionth of a meter), right in the middle of the visible spectrum.

If we pretend that our Earth has no atmosphere and hence no greenhouse effect, the same physics determines the energy radiated as it does for the Sun. For the Sun we know the temperature. For the Earth we calculate the temperature required to radiate the right amount of energy. That is what gives the temperature of –4°F (–20°C, or about 250°kelvin) in the absence of the greenhouse effect. The peak wavelength of the outgoing radiation is in the far infrared at 10 microns.

To calculate the actual effect, we have to know the absorption of our atmosphere at all the relevant wavelengths. Figure 2.1

Technical Note 2.1 (*cont.*)

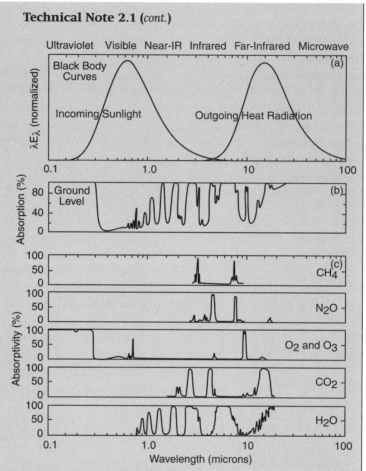

Fig. 2.1(a) Spectrum of incoming sunlight and outgoing heat radiation
for a surface temperature of 60 °F (16 °C). (b) Absorption in percent
versus wavelength for the sum of all the gases shown in the panels
below with atmospheric concentrations measured in the 1950s.
(c) Absorption spectra for five gases as listed, showing the dominance
of water vapor. (*Source*: J. P. Peixoto and A. H. Oort, *Physics of Climate*, ©
Springer, 1992. Reproduced with kind permission of Springer Science
and Business Media)

shows the absorption versus wavelength for water vapor, carbon
dioxide, oxygen, and ozone, and the sum of them all [2]. Also
included is a rough sketch of the distribution in wavelength of
the incoming and outgoing radiation.

The main gases in our atmosphere are overwhelmingly
oxygen and nitrogen. These absorb very little at the important

wavelengths. The absorption spectra for oxygen and ozone are shown in the figure. Ozone is responsible for the total absorption of ultraviolet radiation below wavelengths of 0.3 micron. The peak of the incoming radiation falls at a wavelength where there is practically no absorption.

The absorption spectra of water vapor and carbon dioxide are complex and cover much of the wavelength range of the outgoing black-body radiation. They block a large part of the spectrum. Note that where the absorption is already 100%, adding more of the gas to the atmosphere cannot change the peak absorption value. What does change is the absorption in regions where it is not at 100%. There, adding more will block more. Part (b) of the figure is the total absorption of all the major greenhouse gases, and is very complex. It does not include other minor greenhouse gases, and the data that go into it are from the 1950s when the carbon dioxide level was lower than it is today.

Calculating the response of the climate to a change in greenhouse gases is by its nature complicated. The absorption spectra are complex and include what are called feedback effects (changing one thing changes something else) that will be discussed later. The direction of change can be calculated relatively simply, but for a precise answer, the biggest computers are required to do it numerically.

3

Climate modeling

3.1 INTRODUCTION

Scientific efforts to understand changes in the Earth's climate extend back into the nineteenth century.[1] Computers did not exist, and all calculations had to be done by hand. It was known back then that water vapor was an essential element in the Earth's energy budget and had to be taken into account in any attempt to calculate what the climate might do. Carbon dioxide, one of several known greenhouse gases, was recognized as being particularly important because of its abundance and because it blocks outgoing radiation in some wavelengths where water vapor does not.

There was little concern about climate change until the 1950s when two things happened to wake us all up. Roger Revelle, then Director of the Scripps Institute of Oceanography in San Diego, California, calculated that seawater could only absorb carbon dioxide at one-tenth the rate that scientists had thought, and Charles David Keeling showed that CO_2 in the atmosphere was increasing faster than anyone had thought possible, a finding that agreed with Revelle's analysis. This was a double hit: CO_2 stayed in the atmosphere longer and its concentration was going up faster than scientists had believed possible. It is worth a bit of time to tell the story of how something so important to today's discussions could have been hidden for so long.

3.2 THE FIRST CLIMATE MODELS

The first serious attempt to understand climate in terms of the interaction of the Earth's energy budget with the contents of the

[1] Spencer Weart has posted a splendid detailed history of climate modeling at www.aip.org/history/climate/. He has also documented the history of the

atmosphere was made in 1896 by Swedish chemist Svante Arrhenius, who went on to win the Nobel Prize in Chemistry in 1903 for other work. His much simplified climate model took into account the greenhouse effect, including CO_2. He calculated that reducing the CO_2 in the atmosphere by half would lower the global temperature by about 8 °F (5 °C), which is as much as it was actually reduced in the last ice age. He also calculated that the temperature would increase by about +8 °F (+5 °C) if the CO_2 in the atmosphere were doubled, not very different from today's far more sophisticated models.

This brilliant work introduced for the first time the notion of feedback loops into the discussion of the atmosphere and climate. For example, if the temperature goes up because CO_2 goes up, the amount of water vapor in the atmosphere goes up too, just as vapor coming off a pond increases as the temperature rises. Since water vapor itself is a greenhouse gas, the temperature goes up still more. His model did not include another feedback loop that moves in the opposite direction, the effects of clouds. If water vapor goes up, clouds would be expected to increase. White clouds reflect more of the incoming solar radiation back into space before it gets to the ground than would be reflected by the darker ground itself, and that would decrease the temperature. There are many more feedback loops, and getting them all correct is the main effort of today's climate modelers.

Arrhenius' colleague, Arvid Hogbom, was interested in the entire planetary carbon cycle: where carbon came from and where it went. He thought that human activity, mainly the use of fossil fuels (back then it was chiefly coal), was adding to the CO_2 in the atmosphere. Hogbom and Arrhenius estimated that human activity at that time was increasing CO_2 concentration by a small amount per year (the first estimate was made by Hogbom sometime around 1895). At the rate it was increasing then, it would have taken a thousand years to double the level, which did not seem enough to worry about. They did not conceive of the enormous increase in human economic activity that would come with a booming population and a burgeoning world economy. The amount of CO_2 being added to the atmosphere each year has increased dramatically and the corresponding time to double its concentration has dropped equally dramatically. After Arrhenius, human-induced climate change disappeared from the main scientific

1950s revolution led by Revelle and Keeling at www.aip.org/history/climate/Revelle.htm

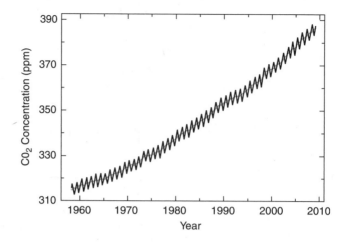

Fig. 3.1   Data from the observatory at Mauna Loa, Hawaii. Atmospheric CO_2 concentration versus year, including Keeling's and more recent observations. (*Source*: Pieter Tans, NOAA Earth System Research Laboratory, Boulder, CO, and Ralph Keeling, Scripps Institution of Oceanography, UC San Diego, La Jolla, CA)

radar screen for reasons that are not clear, but probably stemmed from the belief that change was very slow. While some still wrote about it during the subsequent years, they were generally ignored.

3.3   CLIMATE CHANGE GOES BIG TIME

Carbon dioxide and climate change reappeared with a vengeance when Charles David Keeling of the Scripps Institute of Oceanography (he had been recruited by Revelle because of his interest in measuring atmospheric CO_2) began publishing his measurements of the amount of atmospheric CO_2 in the late 1950s. Keeling's data were taken at an observatory built on a mountain in Hawaii, and showed a small saw-toothed oscillation in the amount of CO_2 in the atmosphere super-posed on a generally upward trend.

For the first few years, the saw-toothed oscillations, caused by seasonal variations, prompted considerable excitement. After several years, however, the big news turned out to be the long-term upward trend in concentration of this greenhouse gas [3]. This was no tiny effect, but a major change in one of the important greenhouse gases, a change that would significantly change the Earth's energy balance should it go on long enough. Revelle's work had predicted that the detailed chemistry of the real ocean greatly slowed the rate at which CO_2 could be taken up by salt water compared with previous estimates. His calculation that the old rate was 10 times too fast was proved correct by Keeling's demonstration of a much more rapid increase than expected in atmospheric CO_2.

Keeling's and Revelle's work triggered renewed interest from the science community. Both theoretical and experimental research on greenhouse gases and their effects increased greatly, and the first formal world conference on possible climate change was organized by the World Meteorological Organization (WMO) in 1979. From then until 1985 WMO and the International Council for Science (ICSU) played a coordinating role in a growing international research effort. The United Nations Environmental Program became involved in 1985, and the WMO/ICSU program was absorbed by the UN and became the foundation of the Intergovernmental Panel on Climate Change (IPCC), formally established as a United Nations sponsored organization in 1987. The first IPCC Assessment Report was produced in 1990 and formed the basis for discussions at the first World Climate Summit in 1992. The IPCC has recently produced its fourth Assessment.[2]

What had changed from Hogbom's and Arrhenius' day, when it was thought that it would take 1000 years before anything of concern might happen, were revolutions in public health and economic development. The world population, which numbered just under 1.5 billion in 1896, had grown to 3 billion when Keeling published his data in the late 1950s and to about 6 billion by the year 2000. The rate of population growth was unprecedented. People were having more babies and living longer, and infant mortality was declining all over the world.

At the same time the standard of living in even the least developed areas of the world also grew. Per capita income more than doubled, and more people with more income meant about a tenfold increase in world economic output [4]. It takes energy to power that output, so energy use increased by a similar amount. This tenfold change, plus the increase expected in this century in total energy consumption, is

[2] A history of the IPCC can be found at http://www.ipcc.ch/about/anniversarybrochure.pdf and IPCC reports can be found at http://www.ipcc.ch/. The reports themselves tend to be technical, but each has a summary for policymakers that is readable by the non-specialist.

Table 3.1 *Removal time and percentage contribution to climate forcing of several greenhouse gases in the year 2000*

Agent	Removal time	Approximate contribution
Carbon dioxide	>100 years	60%
Methane	10 years	25%
Tropospheric ozone	50 days	20%
Nitrous oxide	100 years	5%
Fluorocarbons	>1000 years	<1%
Sulfate aerosols	10 days	−30%
Black carbon	10 days	+20%

responsible for changing the 1896 estimate of a 1000-year CO_2 doubling time into today's 50-year estimate.

3.4 THE BIG PROBLEM: LIFECYCLE OF GREENHOUSE GASES

If greenhouse gases could be removed from the atmosphere rapidly, solving the global warming problem would be much easier. We could wait to see how bad things became, change our ways as needed, and have everything return to normal in a short time. Regrettably, it doesn't work that way. The removal time of some of the main greenhouse gases is measured in centuries. If we wait until bad things happen we will have to live with the consequences for a very long time, no matter how hard we try to fix things.

In 2001 the US National Academy of Sciences, at the request of the White House, reviewed the data on greenhouse gas persistence in the atmosphere and the contribution that each of the gases makes to what is called climate forcing. Climate forcing is the technical term that is related to the change in temperature caused by greenhouse gases. Table 3.1 from the NAS report [5] gives the removal time (how long it would take for something to come out of the atmosphere if we stopped adding to it) and the percentage contribution to total climate forcing from 2001 concentrations of each of the main contributors to climate change. A positive contribution means warming while a negative one means cooling. The one negative contributor is sulfate aerosols which contribute to cloud formation, thereby reflecting more of the incoming solar energy back into space.

Carbon dioxide, the main addition to greenhouse gases, contributes 60% of the warming coming from all the greenhouse gases that humankind has added to the atmosphere as of the year 2000.

Carbon dioxide has a removal time of more than 100 years, perhaps as long as 1000 years. However, only about two-thirds come out in one removal time. The elapse of each removal time reduces the amount remaining by roughly two-thirds of what was there at the start of the period.

Carbon dioxide in the atmosphere has increased by about 40% since the start of the industrial age. Today it is about 380 parts per million (ppm), compared with 270 ppm in the eighteenth century. Sometimes you will see mention of something called carbon dioxide equivalent (CO_2e). This is the amount of CO_2 that would mimic the effect of all of the greenhouse gases taken together. Today's CO_2e is about 430 ppm. (Technical Note 3.1 discusses the sources of the other greenhouse gases.)

There is beginning to be talk of an easy way out of global warming called "geoengineering," where something new is done to cancel the effect of increasing greenhouse gas. The advocates want to introduce another big effect on the climate that no one fully understands, to counteract the temperature increase from greenhouse gases that we also do not fully understand even after 50 years of intense work. To some of my friends who think we should begin work on geoengineering, I have said that it would be unwise because large-scale technical intervention in the climate system can have large-scale unintended consequences (I really said they are out of their minds); it is not smart to count on introducing new effects you don't fully understand to cancel another effect that you do not fully understand. Doing two dumb things rarely gives a smart result. (Technical Note 3.2 has more on geoengineering.)

3.5 THE GLOBAL CARBON CYCLE

The global carbon cycle as sketched in Figure 3.2 tracks where all the carbon comes from and where it goes. The ocean is the largest CO_2 reservoir, holding about 40 000 gigatonnes. (A gigatonne – abbreviated Gt – is the international term for one billion metric tonnes. Each metric tonne is 1000 kilograms or about 2200 pounds.) Next is the land at about 2000 Gt. The smallest of the reservoirs is the atmosphere with about 750 Gt. The land and the oceans take carbon out of the atmosphere with one mechanism and put it back into the atmosphere with another.

Global carbon emissions into the atmosphere in the year 2007 were about 7 Gt per year, but only half of that stays there and becomes

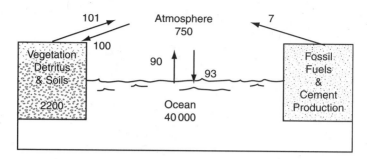

Fig. 3.2   Global carbon cycle. Fluxes (approximate) and amounts stored in the three carbon reservoirs in gigatonnes of carbon.

the main concern for the greenhouse modelers. The rest is absorbed by the land and oceans. Even the part absorbed by the ocean may cause serious problems. The upper layers of the ocean are becoming more acid, with uncertain consequences.

Plants take carbon dioxide out of the atmosphere when they grow, mostly in the spring and summer, and return most of it to the atmosphere when they decay, mostly in the fall and winter. Roughly 100 Gt go back and forth each year. The ocean dissolves carbon dioxide near the surface, and transports it slowly into the deep oceans through the slow natural circulation process. There is also a biological transport in the oceans through the birth of plankton which absorb CO_2 from the air and their death when the CO_2 incorporated in their structure sinks into the depths. Natural evaporation of water vapor at the ocean surface brings dissolved gases including CO_2 back into the atmosphere. Roughly 3.5 Gt more carbon goes into the ocean than comes out, absorbing part of the increased emissions from human activity.

The CO_2 that stays in the atmosphere increases the greenhouse effect. The temperature will increase until a new higher temperature balance is established with enhanced plant growth on land and increased ocean take-up. It is a complicated story: as the surface layers of the ocean absorb more CO_2, they become more acid and less able to absorb CO_2. As the temperature of the water increases, the solubility of CO_2 decreases [6].

There is always a balance between sources and sinks of greenhouse gases that goes to determine the global average temperature. That balance can change as geological processes slowly change. At the end of the last ice age about 15 000 years ago, the balance established between the

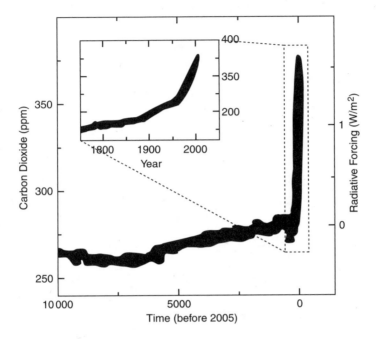

Fig. 3.3   IPCC data on carbon dioxide concentration. Carbon dioxide concentration in the atmosphere from 2005 going back 10 000 years, from ice core and modern data. (Adapted from *Climate Change 2007: The Physical Science Basis.* Working Group I Contribution to the Fourth Assessment Report of the Intergovernmental Panel on Climate Change. Figures SPM.1 and SPM.5. © Cambridge University Press, 2007)

land and ocean sources and sinks of greenhouse gases reflected an average temperature that has changed little since then. That balance gave a CO_2 concentration in the atmosphere of about 270 ppm that lasted until the dawn of the industrial age and the beginnings of large-scale use of fossil fuel. Some analysts believe that the beginning of agriculture about 6000 years ago began to upset the balance. However, there is no evidence of any significant change in CO_2 concentration back then. Though the effect of land use changes is real, it was not large enough to be significant until our recent population explosion. Now, agriculture and land use changes are estimated to be responsible for about 30% of greenhouse gas emissions. Figure 3.3 shows the changes in CO_2 over the past 10 000 years [7]. The rapid spike upwards starting around 1800 should leave no doubt that industrialization has caused big changes in atmospheric CO_2.

The climate-change models have to deal with the entire carbon cycle including how the coupling of the atmosphere to the land and the oceans works. Many research groups all over the world are working on it.

Technical Note 3.1:   Other greenhouse gases

Methane is the second most important greenhouse gas after carbon dioxide, as shown in Table 3.1. It is the natural gas used in power plants and home heating, and is also produced by some biological processes. It is a much stronger greenhouse gas than CO_2 because it easily absorbs the reradiated heat from the Earth over a broader energy band than CO_2 or water vapor. It has increased in the atmosphere from about 0.5 ppm to about 2 ppm and even at this very low level contributes about 25% to today's climate change. Although its lifetime is only 10 years, its action is so powerful that it has to be carefully controlled. Tropospheric ozone is next on the list. This is not the ozone of the Antarctic "ozone hole," but that produced in the lower atmosphere by the interaction of sunlight on ordinary smog, mainly that from automobiles. The short lifetime of this material means that it does not have time to mix through the entire atmosphere, but mainly exists in plumes downwind from large cities. Even though it is not a "whole Earth" effect, its impact on the average temperature increase is important.

Nitrous oxide is also a more powerful greenhouse gas than carbon dioxide. Its preindustrial era concentration was less than about 0.3 ppm. It has increased by only about 15% to date, but even that is significant. Some is produced naturally, and the increase comes from microbial interaction with the huge amount of nitrogen fertilizers used in farming, and from some chemical processes. Its 100-year lifetime means we have to pay attention to it.

Fluorocarbons have a 1000-year lifetime and could have become a serious problem. However, the Montreal Treaty (final version in 1992) to phase out production of fluorocarbons, which was made to close the Antarctic ozone hole, has dramatically reduced its release into the atmosphere. The concern at the time was about ozone as the screen that prevents intense solar ultraviolet radiation from reaching the ground. The fluorocarbons' role as greenhouse gases was little appreciated then. The contrast between the ease of obtaining an international agreement on fluorocarbons and the difficulty of getting an international agreement on CO_2 is an interesting example of the reluctance of governments to enter into treaties that they do not know how to implement. For fluorocarbons, substitutes were available and the economic implications were small. For greenhouse gases, what to do is not so clear and the economic implications are large.

Aerosols are next on the list. These are not the stuff that propels spray out of cans, but substances made by the interaction of chemicals with water vapor in the atmosphere. There are many different ones, but the largest effect comes from sulfate aerosols, which are produced by the sulfur emitted by coal-fired power plants. There are large uncertainties about the effect of these aerosols, but one thing is clear; their effect on climate is to reduce the temperature. They do this by increasing cloud formation, thereby increasing the reflection of incoming sunlight back into space (the reflected light is called the Earth's albedo). Efforts to stop acid rain are reducing the sulfate aerosols produced in the industrialized countries, but increasing coal use in the developing world means that their actual amount is increasing globally. It is not clear how the balance comes out in the long term.

Black carbon (soot) is also an aerosol, but it has a warming effect. Mainly derived from combustion processes, it increases temperature by absorbing more solar radiation than would normally be absorbed by the surfaces which it coats. There remain large uncertainties about its impact.

Technical Note 3.2: Geoengineering

The idea behind geoengineering is to introduce something that has a cooling effect to balance the warming effect of the accumulation of greenhouse gases in the atmosphere. Table 3.1 shows that sulfate aerosols have such an effect. Three main schemes have been discussed. The first is a giant sunshade in space. To reduce the temperature by about 9 °F (5 °C), it would

Technical Note 3.2 (*cont.*)

have to be about 2000 miles across. This would be very good for the aerospace industry.

A second idea is to seed the ocean with the minerals whose lack limits the growth of plankton. There are parts of the world's oceans that are low in the iron needed to support the growth of these tiny creatures. The theory is that supplying the iron will lead to a huge increase in plankton growth; the plankton suck CO_2 out of the atmosphere to build their structure as they grow, and transport the incorporated carbon to the bottom of the ocean when they die, thereby moving CO_2 from the atmosphere to the ocean depths.

A third is to put more sulfate aerosols back into the atmosphere where they will increase clouds and cut the incoming radiation, the sulfur eventually falling out as sulfuric acid and increasing the acid rain we have done so much to cut back. Since these aerosols have only a short residence time in the atmosphere, they have to be continuously added if global warming is to be canceled, and will continually fall out as acid rain.

New geoengineering schemes are being invented all the time. The problem with all of these is unintended consequences. There are thousands of people trying to understand climate change including all the complex feedback effects that make the problem so complex. There are even more people trying to cut back on the emissions that cause it. The geoengineering advocates are prepared to start action without understanding what all the consequences will be. There has already been one experiment at seeding the ocean with iron. It did not work as hoped, but there is nothing to prevent larger-scale tries regardless of our ignorance of the consequences.

The geoengineering folks are starting to back off a bit, and the latest thing is to look at what might be done to counteract a sudden climate instability that caused a rapid, large temperature rise (these instabilities have happened and are discussed in the next chapter). At least the advocates are now talking of doing some serious work on consequences as well as methodology.

4

The past as proxy for the future

4.1 A SHORT TOUR THROUGH 4.5 BILLION YEARS

The global warming debate is about what will happen in the next few hundred years. Our planet Earth is 4.5 billion years old, and over the planet's lifetime changes in temperature, greenhouse gas concentration, and sea level have occurred that dwarf any of the changes being discussed now. Life is thought to have begun roughly 3.5 billion years ago, perhaps earlier, with bacteria-like organisms whose fossils have been found and dated.[1] They lived in the oceans in a world with only traces of or perhaps even no oxygen in its atmosphere. It was about 2.5 billion years ago that the first algae capable of photosynthesis started putting oxygen into the atmosphere, but to a level of only about 1% compared with the 20% of today. All the creatures of the time were small. This earliest period is largely a mystery that is still being unraveled. Recent work indicates that it was only about 540 million years ago that the oxygen concentration in the atmosphere rose to anything like today's values[2] and larger plants and animals appeared.

From then to now saw the rise of many diversified life forms: the growth of giant plants and trees in the Carboniferous era 300 million years ago whose decay and burial gave us the supply of the fossil fuel we use today; a mass extinction about 250 million years ago whose cause is not understood; the rise and disappearance of the dinosaurs in another mass extinction about 65 million years ago, thought to have been caused by the collision of a giant meteor with the Earth. Life is

---

[1] The University of California's Museum of Paleontology has an excellent interactive section on the history of the Earth and the rise of life. See http://www.ucmp. berkeley.edu/exhibits/index.php

[2] A more technical but readable article on the rise of oxygen is by Don Canfield et al. [8].

old; we are young. *Homo habilis*, thought to be our African first ancestor, lived about 4 million years ago. Our particular subspecies, *Homo sapiens*, is only about 100 000 years old. Our civilization is a mere 10 000 years old, a time period so short as to be only the blink of a geological eye. Yet in that eye blink our numbers and economic activity have begun to have effects on a global scale. If we continue increasing emissions at the rate we are now, these effects will become of a size comparable to major geological effects.

The earliest reliable data showing a correlation between temperature and greenhouse gases come from material 55 million years old. In an era called the Eocene, there was a rise in global temperature that seems to have reached a maximum of about 21 °F (12 °C) higher than it is now. This was accompanied by an atmospheric CO_2 level that recent work puts at about 1300 ppm at least, about five times the preindustrial level of the 1700s and perhaps much higher. There may have been other greenhouse gases that increased sharply at the same time contributing to the temperature increase as well. This temperature rise lasted about 100 000 years before the oceans absorbed whatever caused the rise. Though our world was very different back then (no ice in either the Arctic or the Antarctic, the continents not in today's positions, and the ocean circulation very different from that of today) there is an important lesson for us. When the temperature goes up it can take a long time for natural processes to bring it down again, something that will be discussed later.

4.2 THE PAST 400 000 YEARS

There are good data on climate from the ice cores collected by an international team at the Russian Vostok scientific base in Antarctica. These Vostok ice cores let us look at ice laid down earlier than any of the cores from Greenland, for example. They were collected by core drills that brought up miles of core samples and carefully kept them cold for analysis. The Antarctic ice is laid down in well-defined layers as shown in Figure 4.1. Even at the poles, snowfall is seasonal and the bubbles trapped in the ice clearly mark the years. The layers can be counted to date them, and if you want to see what is in the ice from 400 000 years ago, you have to count back through 400 000 layers. Fortunately, there are machines that can do the counting.

Gas trapped in bubbles in each layer can be directly analyzed to determine the amount of CO_2 in the atmosphere at the time the ice was laid down. The temperature can be found by looking at the ratio

Fig. 4.1   Ice-core sample that shows annual banding. This photograph shows a section of the GISP2 ice core from 1837–1838 meters in which annual layers are clearly visible. The layers result from differences in the size of snow crystals deposited in winter versus summer, and resulting variations in the abundance and size of air bubbles trapped in the ice. (Photo from Eric Cravens, Assistant Curator, National Ice Core Laboratory. Credits: NSF, NICL. http://www.ncdc.noaa.gov/paleo/icecore/ antarctica/vostok/vostok.html)

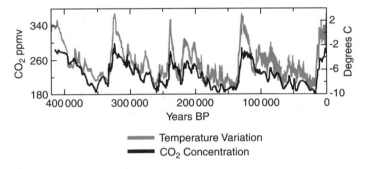

Fig. 4.2   Vostok ice core data. Carbon dioxide concentration in parts per million by volume (left scale) and temperature in degrees centigrade (right scale) relative to today's average temperature back through the last four ice ages. (*Source*: J. R. Petit *et al.* Climate and atmospheric history of the past 420 000 years from the Vostok ice core, Antarctica. *Nature* **399**, 429–436, 1999).

of two stable isotopes of oxygen, ^{16}O and ^{18}O, in the water that formed the ice. The heavier ^{18}O amounts to only about 0.4% of the oxygen in the air. Because of the difference in their mass, the two forms behave slightly differently on condensing into water from the vapor phase, or when evaporating from the liquid. From the measured ratio of the two the temperature can be determined with precision.[3] Figure 4.2 shows the temperature and CO_2 data revealed by the Vostok ice core.

The cores show a fascinating story of repeated ice ages that happened roughly every 100 000 years. They are triggered by changes in the shape of the Earth's orbit from circular to slightly elliptical (100 000 year

[3] The website http://palaeo.gly.bris.ac.uk/communication/Willson/isotopeevidence. html is a good source explaining how this works.

period), the wobble of the poles around the axis of rotation (22 000-year period) and the tilt of the Earth's axis (41 000-year period). The main effect is from the change in the shape of the orbit, but the other two can move the times of temperature minima and maxima around a bit. During the entire 400 000-year record of the cores, the temperature relative to today varied from a high of +3 °C (+5 °F) to a low of –8 °C (–14 °F), while the greenhouse gas concentration varied from a high of 300 ppm to a low of 190 ppm. Carbon dioxide was at 270 ppm at the start of the industrial age and is now at 380 ppm, higher than it has been for at least half a million years.

The temperature cycles seem to cool slowly and warm rapidly. There are occasional spikes of abrupt warming and cooling, the most dramatic occurring about 240 000 years ago. The spikes are not understood, but their existence is the source of concern about possible sudden climate instabilities. Since the main cycles have about the same period as orbit changes they have to be driven by orbit changes, not greenhouse gas changes. It is most likely that the greenhouse gases play a role in the relatively rapid warming through the feedback effects mentioned earlier (in which higher temperature increases water vapor in the atmosphere, increasing temperature further; higher temperature decreases the amount of CO_2 that can be dissolved in the surface layer of the ocean, increasing the temperature further, etc.). The temperature spikes occur in both the warming phases and the cooling phases, indicating instabilities that we do not understand.

What all of this tells us is that greenhouse gases and climate are coupled. Throughout 400 000 years and many ice ages, the greenhouse gas concentration has never been as high as it is today.

4.3 THE RECENT PAST

The temperature record for the past thousand years is called the "hockey stick" curve because it looks like one: roughly flat except for the past hundred years, when it swoops up like the end of a hockey stick. However, the thermometer was only invented in 1714, and reliable worldwide records exist only for the past 150 years. Everything before that comes from what are called proxies for the temperature – other things that are indirectly related to the temperature. Unfortunately, all of the proxies are affected by more than just the temperature. Tree rings are one example of such a proxy. The width of the rings does depend on the temperature, but also depends on the amount of rainfall in the growing season, so the temperature cannot be accurately determined by tree rings alone. Many proxies are used together in the hope that they will give a reasonable average. Here are a few. The date of the grape harvest at wineries is a proxy for the heat of summer: the earlier the harvest, the hotter the

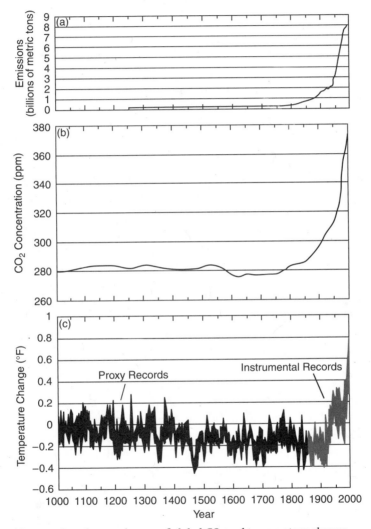

Fig. 4.3   One thousand years of global CO_2 and temperature change. (a) Estimate of past emissions. (b) CO_2 concentration from ice-core data. (c) Temperature relative to the 1000-year average from proxies to about 1850 and from instrumental records to the present. (*Source*: Climate change impacts on the United States, http://www.usgcrp.gov/usgcrp/Library/nationalassessment/foundation.htm)

summer. The length of glaciers serves to tell us something about year to year temperature changes. The layer thickness of corals laid down each year depends on the temperatures of the shallow waters in the coral reef.

Figure 4.3 is from a report published in 2001 titled "Climate Change Impacts on the United States." The report analyzed what would happen in the United States if the temperature went up, and the figure is used here to show what we know about the correlation of temperature, greenhouse gas concentration, and economic activity. I was a member of the review panel for this report, and our job was to ensure that the report was based on good science and that uncertainties in its conclusions were clearly identified. This review marked my first direct involvement in climate-change impact studies. One of the strengths of the report comes from its use of two climate scenarios, one toward the lower end of today's estimates of temperature rise and one toward the upper end. It makes interesting reading even today if one wants to know what will happen in the United States as the temperature rises.

Figure 4.3(c) shows the temperature data averaging all proxies from 1000 years ago up the beginning of the instrumental record and from the instrument record itself for the past 150 or so years. Figure 4.3(b) shows CO_2 concentration. These data are from ice cores and are solid. Figure 4.3(a) shows carbon emissions into the atmosphere. Here, the curve is an estimate based on economic activity (see Ref. [4]). It is not very precise, but it doesn't have to be. The correlation between the three panels is remarkable.

Those who deny the reality of global warming argue that the apparent flatness of the temperature record before about 1800 is wrong and underestimated the climate's natural variability. The National Academy of Sciences (NAS) put together a panel to examine the issue. Their report showed a much larger variability from the proxies than had been claimed by some. It contains a detailed analysis of how the past temperature can be reconstructed from a multitude of proxies.

Results from the National Academy analysis are shown in Figure 4.4. We are interested in average temperatures so we need proxies distributed over a wide area of our world. Tree rings are better than the date of the wine harvest because we can get tree records from all over, whereas the wine records are mainly restricted to Europe. The uncertainties in the temperature derived from each proxy increase as you go further back from the present.

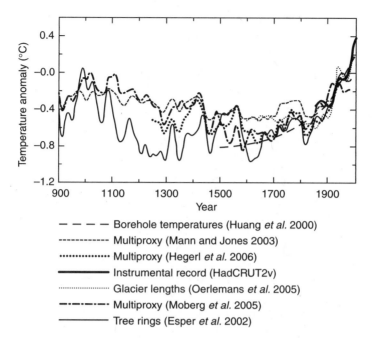

Fig. 4.4   Proxy and instrument data from the NAS Report. NAS
analysis of temperature for the past 1200 years from six proxies
(the reference citations are those in the original report). The heavy,
black line is from the instrument record. (Reprinted with permission
from *Surface Temperature Reconstructions for the Last 2000 Years*, © 2006
National Research Council. Courtesy of the National Academy Press,
Washington, DC. www.nap.edu/catalog/11676.html Figure S-1)

Nothing in the past 1200 years is like the sharp increase in temperature that began in the nineteenth century coinciding with the
increase in the use of fossil fuels. Natural processes do not normally
change the global average temperature this fast. The most likely cause
is human activity.

5

Predicting the future

5.1 WHO DOES IT?

There are many sayings about the difficulty of predicting the future. My favorite is, "Predicting the future is hard to do because it hasn't happened yet." It is especially hard when you are trying to predict what will happen 100 years from now and the science behind the prediction is really only 50 years old. It was the work of Keeling and Revelle in the 1950s mentioned earlier that jump-started the science community's work on climate change and global warming. It is the Intergovernmental Panel on Climate Change (IPCC) that does the predictions today.

My own involvement in climate change research has been more as an observer than as a participant. My first exposure to the issue was in 1978 when a group I am in, called the JASONs, took it up. The JASONs are a collection of academics mostly that meet every summer for about six weeks to work on problems of importance to the government. In 1978 a subgroup of the JASONs led by Gordon MacDonald, a distinguished geophysicist, began a study of climate change for the US Department of Energy. The JASONs always have many pots on the stove and I was working on something else. However, we all were fascinated by the climate issue, and nearly everyone sat in on the sessions and critiqued the report. Its conclusion was that doubling atmospheric CO_2 would increase the average global surface temperature by 4.3 °F (2.4 °C), and that the increase at the poles would be much more than the average. The JASON climate model included a more sophisticated treatment of the ocean–atmosphere interaction than had been used before. The model was a simplified one that could be solved without big computers, and the answer was in fairly good agreement with what we get now for the average temperature increase, but overstated the polar increase. The report was influential in increasing govern-ment funding for climate change research.

The IPCC, which shared the Nobel Peace Prize in 2007 for its work, was created in 1988 as a UN-sponsored organization under the United Nations Environmental Program.[1] With the signing of the UN Framework Convention on Climate Change (UNFCCC) in 1992, the IPCC became the technical arm of this 162-nation organization.

The IPCC does not sponsor climate research itself but coordinates and synthesizes the work done by many groups around the world. Its major products are its periodic Assessment Reports which make the best scientific predictions of the effects of increasing greenhouse gas concentration to the year 2100 based on the state of knowledge at the time the report is produced. The first Assessment Report appeared in 1990, the second in 1995, the third in 2001, and the fourth in 2007.

The first IPCC assessment was due only two years after the cre-ation of the organization, and required a huge amount of work to produce in so short a time. The team had to gather and analyze the data, create a credible scientific peer-review system, and get the report through its parent UN agencies and out to the world. The report on the science of climate change (IPCC Working Group I or WG I) was broadly accepted in the science community. The report on expected impacts of climate change (WG II) encountered some scientific argument, while the report on responses (WG III) wandered into the policy area and ran into serious troubles with the IPCC's UN sponsor who thought that policy was their job. The policy part was removed to the UNFCCC organization itself, and the IPCC remains today as the main organiza-tion responsible for scientific and technical analysis of the issues. It is respected by governments, non-governmental organizations, and the science community. The process of producing these reports is compli-cated but the output of the IPCC has come to be trusted by all the sign-ers of the UNFCCC, which means most of the members of the UN.

After 1992 and the signing of the UNFCCC, more formality was brought to the assessment process. The assessments are now prepared by a large group of experts who are nominated by signatory countries. The only way, for example, that a US scientist can become a mem-ber of an assessment team is by nomination by the US government or some other country (any signer country can nominate anyone).

---

[1] All the IPCC reports are available online (www.ipcc.ch) and the reports called "Summary for Policymakers" are written with clarity for people without a techni-cal background.

Although the administration of US President G. W. Bush was not noted for believing in the urgency of action on climate change, it did nominate the best climate scientists in the United States to the relevant panels. According to the panel members that I know, there seems to have been little politics in the selection of the scientific members of these panels by any country.

Each of the three Working Groups produces what is called a Summary for Policymakers. This summary is non-technical and is gone over line by line with representatives of the signers of the UNFCCC at a meeting that has political as well as technical overtones. Since the UN operates by consensus on climate change there has to be agreement on the exact wording of the report, and that language evolves as the scientific evidence evolves. For example, in the Third Assessment Report the summary did not say that global warming was caused with high probability by human activities. The Fourth Assessment Report does say that. There was no consensus in the Third Report that human activities are the main cause of warming (the main holdouts were China and the United States), but there is in the fourth after a long and sometimes heated argument at the review meeting.

After agreement is reached on the wording in the summary, the scientific groups have to go back and make the words in their technical reports consistent with what is in the summary. They may have to change their descriptive words but they do not have to change their technical findings or any of the numbers in their analyses. Some would say that this procedure is overtly political. They would be correct, but since only the countries that are the major emitters of greenhouse gases can do anything about global warming, a consensus on the issues is needed as a preface to global action. Without that consensus the two largest emitters of greenhouse gases, China and the United States, have refused to join in official control mechanisms. With it, they may join in the next round of negotiations on international action.

5.2 HOW IS IT DONE?

All sorts of models are made of what will happen in the future, based on previous experience and knowledge of the processes that will affect whatever is being modeled. People make (or should make) models of income and savings against payments when buying a car. The Federal Reserve models economic growth and inflation when it decides on interest rates. Both of these are relatively short-term models that rely on predictions that look only months or a few years into the future, and they are also not based on any actual physical laws. They also do not treat the potential for instability in any reasonable way, as shown by the global financial chaos that began in 2008.

The climate models are attempting to do something much more ambitious, and do it better. They are trying to predict what will happen to our climate 100 or more years in the future based on models of how the climate system responds to changes. The models are grounded in the physical and biological sciences, are mathematically complex, and come from an evolving understanding of the science. In Chapter 2 an introduction to the greenhouse effect was given that showed how the temperature of our planet is set by a balance between the energy coming from the Sun and the heat energy radiated back out into space. Calculating the temperature at which that balance is struck in the real world is an enormously complicated job that has to take into account a host of interactions between very many elements of the systems that determine our climate.

The amount of incoming radiation from the Sun is known very well. Not all of that radiation reaches the ground. Some is reflected back into space by clouds, and some is absorbed by various chemicals in the atmosphere and by the clouds themselves. Some is reflected from the surface, more strongly from snow and ice, less strongly from deserts, and least strongly from the oceans and land areas covered by vegetation. The oceans, the land masses of the continents, and the atmosphere interact with each other in complex ways. Changes in one thing change other things as well. These effects are called feedback loops and some were described earlier. I have mentioned how increasing greenhouse gases increases the temperature; increasing the temperature increases water vapor in the atmosphere; water vapor is also a greenhouse gas so the temperature increases further; increasing water vapor also increases clouds; more clouds reflect more incoming radiation into space, decreasing the temperature. We usually think that increased temperature should lead to less snowfall, but one of the oddest feedback effects increases snowfall in Antarctica when the global average temperature increases slightly, because an increase in temperature increases water vapor in the atmosphere and that leads to more snow. Some of these feedback loops amplify climate change effects while others reduce them. Getting all of this right is the job of the modelers and their computers.

Many of the feedbacks are positive in the sense that given a temperature increase they increase the temperature more, but we know that there has been no runaway greenhouse effect on Earth where the feedbacks turned the Earth into something like Venus. The historical record over the past few billion years shows that although the climate has been both much hotter and much colder than today, it has stayed within limits that still support life. If the Earth were to become either very hot or very cold compared with today it might not support our standard of living, but life would go on.

There are many climate models that have been created by groups of experts around the world. The most sophisticated and complex are called atmosphere–ocean general circulation models (AOGCM). These models divide the surface of our planet into small blocks, the atmosphere into layers, and the oceans into layers too. Here the first big problem arises. The surface of the land is not smooth. The wind in the atmosphere is not uniform. The ocean has currents that are narrow compared with the size of the ocean. In size, the Gulf Stream that keeps Europe warm is to the Atlantic Ocean as the Mississippi River is to the North American continent. The Alps are small compared with the total land area of Europe. All are small compared with the scale of what they affect. Models have to look at effects at an appropriate scale. On the land surface, mountain ranges affect wind patterns and therefore the transport of heat. In the oceans, currents such as the warm Gulf Stream off the US east coast and the cold Humboldt Current off the US west coast pierce the quiet oceans with water plumes that also move huge amounts of heat. Land bottlenecks exist around Greenland, for example, that restrict the flow of water, strongly affecting the heat flow.

All of these effects make the design of the calculations extremely complicated, and the description above only begins to take into account the wide variety of conditions over the entire surface of the Earth. The most sophisticated calculations start off with the surface divided into squares which might be as small as 25 miles on an edge where the terrain is highly variable and as large as several hundred miles on an edge where the terrain is smooth and fairly uniform. The atmosphere is divided into layers and there can be as many as 15 or 20 of them. Similarly the oceans are divided into layers, and the number of layers has to take into account the depth of the oceans as well. In the actual calculations there can be hundreds of thousands of these cells. Heat and fluids (water in the oceans and air in the atmosphere) flow into a cell from one side or top or bottom, and flow out another to adjacent cells. The calculations require enormous computers, and even the largest computers available today cannot do the job quickly. These AOGCMs are not run very often because of the huge amounts of computer time needed. There are 23 different AOGCMs that the IPCC takes into account in its climate synthesis.

The problem is so large that it cannot be solved from first principles on any existing computer. That would require starting off with oceans uniformly full of water and an atmosphere full of the proper mix of gases both at the average temperature of the Earth (65 °F or 18 °C), and running the program for long enough to allow the currents in the ocean and the air to develop, the temperature to stabilize, the ice caps to form, etc. Perhaps when computers become at least 1000 times more powerful than those of today it can be done. Today, what is done is called a "perturbation analysis." The starting point is the world as it is, and the calculation sees how it changes (is perturbed) when greenhouse gases are added to the atmosphere. The oceans are there, their currents are flowing, the atmospheric winds blow, the ice is in place, and the computer grinds away step by step to predict the future as greenhouse gases accumulate.

There are also natural phenomena that occur randomly from time to time and have to be put in explicitly. For example, major volcanic eruptions, like that of Mount Pinatubo in the Philippines in 1991, throw large amounts of material into the upper atmosphere that affect the albedo (the reflection of incoming solar radiation back out into space) directly and indirectly by affecting cloud formation. This gives a cooling effect that lasts a few years until the volcanic material falls out of the atmosphere. Things like this cannot be predicted in advance. After an event like Mount Pinatubo, the material ejected into the sky has to be added explicitly and the models run again. Fortunately for those doing the predictions, effects from these kinds of events do not last for a long time and are not really important for the long term, although they do contribute to the seemingly random fluctuations in the planetary temperature.

A more important issue is predicting how human activities will change the amount of greenhouse gases put into the atmosphere. Scenarios are created that predict how energy use grows over time and what the mix of fuels will be. From this the amount of greenhouse gas going into the atmosphere for each scenario is derived. The IPCC uses six main scenarios, each with a few variations. These scenarios go through the same sort of approval process as the climate change reports to assure the UNFCCC signatories that the scenarios are reasonable. The scenarios do not assume the existence of any mechanisms for greenhouse gas reduction; the IPCC is not allowed to make such assumptions. The scenarios are simply alternative economic growth models that make different assumptions on economic growth, energy efficiency, population, fuel mix, etc. For example, all the scenarios assume a world economic growth rate of between 2% and 3% per year. This does not seem to be much of a difference in economic growth in the short term, but over a period of 100 years that 1% extra economic growth makes world economic output, energy use, and greenhouse gas emissions more than twice what they would be with a growth rate of only 2%.

The calculations move ahead one time step at a time. The model adds a year's worth of greenhouse gases to the atmosphere and calculates what happens to the atmosphere, the oceans, the clouds, the snow and ice, etc. This answer is the input for the next time step and so forth on into the future.

5.3 RESULTS

The IPCC cannot yet predict a specific number for the temperature increase by the year 2100 for each of its scenarios. What they can do is to predict for each scenario a rough upper and lower limit or range of the temperature increase, and an average value. The actual increase for each scenario can lie anywhere in its range with about a 70% probability. The range of outcomes comes from the use of many climate models that are produced by independent groups. Climate change science is still evolving, and different groups give different weights to different phenomena. To me this is one of the strengths of the system. The independently developed collection of models from many groups minimizes the risk that a not-well-founded majority view would dominate the analysis.

Figure 5.1 gives some of the results of the most recent IPCC analysis. The graph shows the historic data from the twentieth century (heavy black dotted line) and the average of the model predictions for three of the scenarios for the twenty-first century. The range of the prediction from the many models used in all of the IPCC scenarios is given in Table 5.1. The scenario A1FI is closest to "business as usual" (continuing on with the same mix of fuel as we use today). It also gives the largest temperature rise. The range of predicted temperature rise is large for all of the scenarios. In the A1FI scenario it is so big as to span a range from merely difficult to live with to very disruptive to society. Narrowing the range of outcomes for any of the scenarios requires sorting out which of the many models is most nearly correct. As we shall see, it will take about 30 years for the

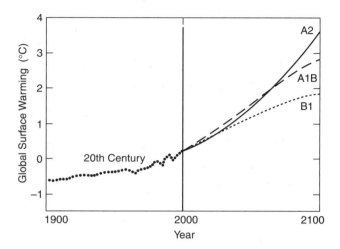

Fig. 5.1   IPCC projections of twenty-first-century temperature increases. The best estimate average temperature change from the year 2000 to 2100 relative to the year 2000 predicted for three of the IPCC scenarios from the collection of different models as shown in the Fourth Assessment Report. The solid line is for scenario A2, the dashed line for A1B, and the dotted line for B1. Also shown are the historic data for the twentieth century. (Adapted from *Climate Change 2007: The Physical Science Basis. Working Group I Contribution to the Fourth Assessment Report of the Intergovernmental Panel on Climate Change*, Figures SPM.1 and SPM.5. Cambridge University Press)

temperature to change enough to show which of the models does the best job. If it turns out that the predictions giving the larger temperature rise are the correct ones, it will be much more difficult to limit the ultimate temperature increase if we do nothing until we know the answer.

All of the models are quite complex and have adjustable parameters that I call "knobs." These knobs can be turned to increase or decrease the importance of some of the processes that go into making up the model, and each model turns its knobs to get agreement with the past. I call this tune-up process "postdiction," fixing up things so that you agree with what you know has already happened. What we really want to see is the best prediction of what will happen in the future, but we cannot know which of the many models agrees best with future facts until enough time has passed to collect the necessary data to show the differences between reality and model prediction. Remember that all of the models say the temperature is going up; they differ by how much it will go up.

Table 5.1 *Projected global average surface warming at the end of the twenty-first century*

| Case | Global average temperature changes relative to 1980–1999 | | | |
| | Best estimate | | Range | |
	°F	°C	°F	°C
B1	3.2	1.8	2.0–5.2	1.1–2.9
A1T	4.3	2.4	2.5–6.8	1.4–3.8
B2	4.3	2.4	2.5–6.8	1.4–3.8
A1B	5.0	2.8	3.1–7.9	1.7–4.4
A2	6.1	3.4	3.6–9.7	2.0–5.4
A1FI	**7.2**	**4.0**	**4.3–11.5**	**2.4–6.4**

I will use the A2 scenario to make an estimate of how long it will take to learn which of the models are most nearly correct. When enough time has passed the models will not need scenarios to compare with the data. They will have the actual change of greenhouse gas concentration over time as their input. Scenarios will still be needed to predict the future under different assumptions on energy use, technology development, population growth, etc., but the spread in the future temperature rise will be reduced because those models that disagree with the observations will be discarded (or more likely, their knobs readjusted).

The complicating factor is noise in the system. The average temperature of the Earth varies randomly from year to year by an average of 0.2 °C (about 0.4 °F) above and below the trend line. Sometimes the random jump is less, sometimes more, but the long-term average noise is quite consistent. Figure 5.2 for scenario A2 shows bands centered on the upper end of the range, the average of the models, and the lower end of the range. Each band is 0.4 °C wide to represent the climate's random jumps around the trend lines. You cannot really tell where the temperature change is headed other than that it is headed up, until the bands begin to separate. What has happened in the past is no help since all of the models are tuned to agree with what has already happened. Their predictions of the future in any of the scenarios are what differ, and it will take about 25 more years to see which of them best agrees with what happens in nature.

The models will continue to evolve and improve as information comes in. The current rapid melting of the Arctic ice was unexpected, for example. Why it is happening will be something better understood

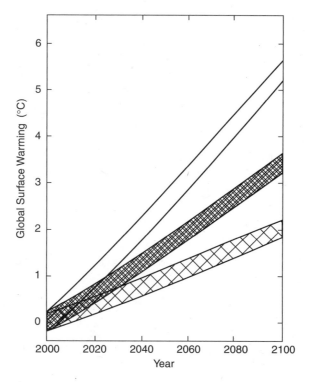

Fig. 5.2   Temperature change estimates including random fluctuations. For scenario A2 the bands represent the upper end of the range, the average of the models, and the lower end of the range broadened into bands by the random jumps of 0.2 °C around the trend lines. Where the temperature change is headed cannot be accurately determined until the bands begin to separate. That will take about 25 more years.

by the time of the next assessment report. The movement of the Greenland glaciers seems to have been misunderstood. Their movement accelerated in 2006 and 2007, but then slowed back to the more normal pace in 2008. That too will be looked at in depth for the next report. Personally, I think there is not enough biology in the analysis. Part of the transport of carbon into the deep ocean is governed by the growth and death of plankton which absorb CO_2 as they grow and transport it down when they die. The effect of changing temperature on the growth of plants needs more work too because plants also contribute to sequestering carbon in the soil. The models will continue to evolve, and each cycle should be a better approximation to the real world than the last. It is important that there continue to be many independent centers working on the problem.

A friend who has been kind enough to read and comment on drafts was worried about this chapter. His concern was that by talking about uncertainties, I make it easy for those who oppose action to argue for more delay. The only advice I can give is to beware of quotes out of context and to beware especially of incomplete quotations. The last two sentences of the paragraph two above can, by selecting only some of the words, be made to say what they did not say. You should be suspicious if you see something like, "Nobel Prize winner says 'You cannot really tell where the temperature change is headed... That will take about 25 more years.'"

A few more words about uncertainty are needed. The range for the temperature rise given in Table 5.1 does not span all the possibilities. It is only the most likely range. The chance that the temperature will go up by an amount outside the range is about 30% according to the IPCC. However, if it is outside the range it is more likely to be on the high side than on the low side.[2] Also, though there is uncertainty, all of the models predict a temperature rise.

5.4 WHERE ARE WE?

This is the end of the climate-change section of this book. I have taken you through the science behind global warming and the methodology used to predict what will happen in various scenarios. There are uncertainties in temperature change expected in each of the scenarios, but the one that people should be concentrating on is the one called business as usual or BAU (the IPCC scenario called A1FI is closest to this). The A1FI scenario run through all of the models predicts an average global best-guess temperature rise of about 7 °F (4 °C) and a range from 4–12 °F (2–6.5 °C), with a rise about twice that at the North Pole. The consequences will be disruptive at the low end and destructive at the high end. If you live in California there will be no snow in the winter, and the water available in the summer will have to come from large dams and reservoirs that have not been built yet. If you visit Florida, you will travel in boats over much of the area where people now live. If you live in the temperate-zone farm belt, you will be growing bananas rather than wheat. If you live anywhere you will worry about the movement north of the diseases and insects that now are in the hotter zones nearer to the equator. More detail is in the report of the

---

[2] If you are mathematically inclined you can see why this is so in an article by G. H. Roe and M. B. Baker [9].

IPCC Working Group II, "Impacts, Adaptation and Vulnerability."[3] You should also note that the actual rate of increase in greenhouse gases in the atmosphere is even faster than that used in the A1FI scenario.

Before moving on, I want to leave you with a word of warning. There are still some who claim that global warming is some kind of hoax or conspiracy by the science community. There is little use in getting into an argument with them because they will not listen. The Summary for Policymakers in the 2007 IPCC report begins: "Warming of the climate system is unequivocal, as is now evident from observations of increases in global average air and ocean temperatures, widespread melting of snow and ice, and rising global average sea level." It goes on to say, "Most of the observed increase in globally averaged temperatures since the mid-20th century is very likely due to the observed increase in anthropogenic GHG concentrations." Earlier I mentioned that the IPCC works by consensus. Every line in a Summary for Policymakers is gone over word by word with representatives of the signers of the UNFCCC. The quote above has been agreed to by all of the nations. This is a change in position by both the United States and China from skepticism to agreement with the world consensus.

Beginning with the next chapter, I am going to shift gears to discuss the sources of the increase in greenhouse gases (mainly fossil fuels) and the coupling of energy use to economic growth. The BAU scenario is simply the continued use of our present mix of fuels as the world economy and population grow. I will then go on to discuss what I think our target for the allowable greenhouse gas concentration should be at the end of this century and how we might get there.

While part of the increase in greenhouse gases comes from changes in land use and from some industrial processes, by far the most comes from the energy sector. Energy use is usually divided into three categories: transportation; residential and commercial; and industrial. They all have different problems and decarbonizing each sector requires different solutions. The early steps are easy, but the problem gets harder the deeper the cuts have to be. While we are implementing the easy things, we will have time to improve the systems that can be used in the more difficult parts of the problem. Threaded through this section will be some discussion of policies that can help and of some poor choices that have hurt.

[3] http:// www.IPCC.ch

By now it should be clear to the reader that I think we should begin now to head off the worst effects of global warming. As we will see, it is not possible to return to the preindustrial level of greenhouse gases in this century. It is not even possible to stabilize the atmosphere at today's levels of greenhouse gases in this century. The nations of the world might stabilize the level at twice the preindustrial level, but only if they get to work, and if the largest emitters set an example that the rest of the world can follow. In the United States and the European Union there is much talk about reducing emission to 80% below their 1990 values by the year 2050. That is probably not possible for the rich countries using only today's technology, and is certainly not possible for poor countries that are trying to increase their per capita incomes and move up out of poverty. The rich countries have to develop the needed technologies for all to use, and some accommodation is required between what the rich are expected to do and what the poor are expected to do. Later I will show what a profile of world average emissions versus time has to be to stabilize the atmosphere at some agreed greenhouse gas level (I use double the preindustrial level). When the world agrees on a program, the rich countries will have to do more and the poor countries less initially, while later all will have to do the same thing.

I have two very young granddaughters, and it would not be responsible of me to leave this problem to them. Whatever problem we leave them will take centuries to fix. With apologies to Shakespeare, I rewrite Hamlet's soliloquy as:

> To act, or not to act: that is the question:
> Whether 'tis nobler in the mind to have our children suffer
> The slings and arrows of global warming,
> Or to take arms against this sea of troubles, And by opposing
> end them?

Ending them is what we should be doing. How to act and how fast to act are the next questions.

Part II Energy

6

Taking up arms against this sea of troubles

6.1 INTRODUCTION

This chapter begins the discussion of how we can get out of the climate change trap that the world is in because of economic growth, population growth, and a lack of understanding of how our actions affect our environment. Though even the poorest are better off than they were a century ago, global warming will reverse the improvement in the lives of all, unless we do something about it. The source of the problem is the energy we use to power the world economy, and the agricultural practices we use to feed the world population. The problem is solvable, but the solution requires global action.

All of the major emitters of greenhouse gases have now agreed that the problem is real, but have not agreed on how to share the burden of cleaning things up. It will be hard to devise a system of action that allows the developing nations to continue to improve the welfare of their citizens while they also reduce emissions. The consequences are in the future while action has to begin in the present, and that creates difficult political problems for all nations because the costs are now, whereas the benefits will come later (I come to that in Part III).

I start here outlining the sources of the greenhouse gases that cause the problem, how the projections of future energy use that dominate emissions are made, and how we have to reduce emissions over time to stabilize the atmosphere at some new, not too dangerous level. The longer we wait to start, the harder it will be to solve the problem because the emissions will be larger and reductions will have to be larger, faster, and more expensive. The next chapter is about what the economists have to say about how fast to go in reducing emissions. After that, I move to the specifics about various forms of energy.

There are many different greenhouse gases, some of which have been mentioned already. Each gas has a different contribution to climate change, and the modelers talk in terms of CO_2 equivalents or CO_2e. What the models do is to calculate the effect of emissions of all the greenhouse gases and compare the effect to the emissions of the amount of CO_2 that would produce the same effect, hence the name CO_2 equivalent (see Technical Note 6.1).

About 70% of the anthropogenic emission of greenhouse gases (most of the CO_2 and part of the CH_4) comes from the energy used to generate electricity, make buildings usable, run all transportation systems, and supply all the energy needs of industry. The rest comes mainly from agriculture (fertilizer use is the main source of N_2O), changes in land-use patterns driven by the search for higher crop yields, and deforestation in the search for new lands to grow crops or convert to other uses. The total man-made emissions in the year 2004 according to the IPCC were 49 billion tons per year of CO_2e. The energy sector is both the largest and fastest growing source of emissions and this is what I focus on here, but agriculture and land-use changes are a large problem that I only touch on in a later chapter on biofuels. Biofuels may be a case where the consequences of adopting a self-serving proposal from agribusiness have made things worse instead of better. Agriculture and land use are problems that deserve much more attention than they are getting, but the main focus in the rest of this book is on the larger issue, energy usage.

6.2 ENERGY NOW AND IN THE FUTURE

The use of commercial fuel drives the economies of the world. Countries using the least energy per capita have the least income per capita and their people are the poorest. Countries using the most energy per capita have the largest incomes per capita and their people are the richest. The poor want to grow rich, the rich want to grow richer, and so energy consumption everywhere in the world continues to rise.

The very poorest countries are not now relevant to world energy demand or to the greenhouse gas emissions that drive climate change. There are about 1.6 billion people who have no access to any form of commercial energy. If they were magically given enough to run a refrigerator, light their homes at night, and run their schools, the added energy required would amount to only about 1% of the world's energy consumption. These countries will begin to have an impact on energy demand and climate only when their economies grow enough to make a difference. Until then, they should be left to increase the well-being of their citizens in the most effective way they can without regard to global climate issues. Of course they have to be careful about their local environment, but mandated greenhouse gas reductions should not be required of them.

Predicting future energy demand is done by predicting two other things, economic output (gross domestic product or GDP) and a quantity called energy intensity (E_i, the amount of energy required to produce a unit of GDP). Multiply the two together and you get the energy required to generate that GDP.

$$\text{Energy} = \text{GDP} \times (\text{Energy/GDP}) = \text{GDP} \times E_i$$

For those readers that remember their algebra this looks like an identity: cancel GDP with the GDP in E_i and it only says energy equals energy. It does indeed, but it allows the use of numbers that can be estimated with reasonable confidence for a region, a country, or the world in the prediction of future energy demand. Estimates of both GDP growth and changes in energy intensity are based on historic trends, and with lots of past data there is more confidence in predicting the future by using them rather than just guessing what energy growth will be.

Energy intensity is a measure of efficiency and of the product mix in a particular economy. Energy intensity usually drops as an economy matures, largely because of a shift from manufacturing to services (it takes much less energy to run a bank than a steel mill, though both may produce the same increment of GDP). This is particularly important because two of the world's largest countries by population, China and India, are undergoing rapid economic growth. At the beginning of their growth cycles, industry dominates over services and processes tend to be relatively inefficient. The effects of improving efficiency (reducing energy intensity) on energy demand are also important in estimating the worldwide demand for energy in the future.

Figure 6.1 shows the energy intensity of the US economy going back to the year 1800 when horsepower meant literally the number of horses used. The main sources of power then were animals, wood, and waterwheels. Energy intensity declined at an average rate of about

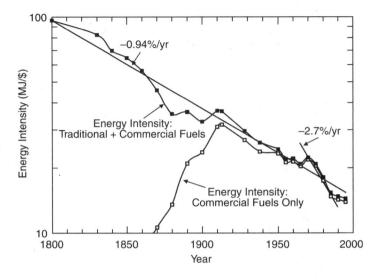

Fig. 6.1   Energy intensity of the US economy in megajoules per dollar
from 1800 to 2000. (Courtesy of Professor S. Fetter, University of
Maryland. © 1999 Steve Fetter, *Climate Change and the Transformation
of World Energy Supply*. Stanford, Center for International Security and
Cooperation)

1% per year, right through the transition to commercial fossil fuels
(coal first and then oil) in the late nineteenth century. There is one
notable exception, the period from about 1970 to 1985. This was the
time of the OPEC oil embargo, and the Iranian revolution. The resulting huge increase in oil prices drove a large-scale and successful effort
to improve the efficiency of energy use throughout the United States
and the world economies. In the mid-1980s, oil prices dropped and
the economies went back to their old ways. The high prices of fossil
fuels today are once again driving a move toward more efficient use of
energy, which is good for the economy and for reducing greenhouse
gas emissions. In the industrialized world, energy intensity is now
declining at a rate of about 1.3% per year. China's most recent 5-year
plan has a goal of reducing energy intensity by 4% per year between
2005 and 2010.

To predict future energy demand, I will use data from a study
by the International Institute of Applied Systems Analysis (IIASA) and
World Energy Council (WEC) [10]. The income and energy-intensity
data are from 1998 and earlier. Although these are not the most
recent data, their report is available on the Web and the viewer can

interactively see what happens as assumptions are changed. The projections are close enough to more recent ones that trends and broad conclusions are unaffected.

Before going on, a word about population is in order. Income and population taken together give a measure of the standard of living, the per capita GDP. Much of the drive for rapid economic growth comes from pressures to increase the standard of living of a country's population. The world population was about 1.5 billion in 1900; it doubled to 3 billion by 1960; it doubled again to 6 billion by 2000. This booming population required an associated boom in energy use to maintain the world average standard of living, and an even larger boom to improve it. If the time for the population to double continued to shrink, the world would be in even more serious trouble than it is today, but, fortunately, population growth is slowing.

Population can be predicted with considerable accuracy for the next 20 to 30 years and with lesser accuracy for the rest of the century. Almost every woman who will have a child in the next 20 years has already been born. We know current fertility factors (children per female) and we can project fertility with confidence for several decades. The longer-term projection is where the major uncertainty lies. Fertility has been declining all over the world, and each new long-term prediction implies it will decline faster than assumed in the last prediction. It seems almost universal that as income goes up, fertility goes down. Some ascribe this to the move to the cities where fewer children are needed to support the family. Others say that it is simply that as women have more opportunity they have fewer children. Whatever the reason, predictions for future population have been coming down over the years.

Population trends for the world are analyzed by the United Nations Population Division (www.un.org/esa/population/unpop.htm), which breaks them down by country and region. Each recent projection has given a lower long-range population than the previous one. In 1998 the world population estimate for 2100 was 10.5 billion while the most recent estimate made in 2004 predicted about 9 billion by 2050, remaining there for the rest of the century. The slowing of population increase eases the pressure for growth and for the use of ever more energy. It is likely that the next UN population projection will predict that the year 2100 population will be even lower. However, the change will not be very large and will have little effect on energy use. All this of course assumes no major war, epidemics, or other worldwide societal disruptions.

Returning to economic growth, the IIASA-WEC economic growth scenario closest to what has been happening is their scenario A (high growth). The long-term growth rates assumed are about 1.6% for the industrialized world and nearly twice that for the developing world. The methodology divides the world into three groups. The "Industrialized" countries are the United States, Canada, Western Europe, Japan, Australia, and New Zealand; "Reforming" countries are the old Soviet Block – Central and Eastern Europe plus all the States coming from the break-up of the Soviet Union; "Developing" countries are all the rest and are dominated by China and India. Under their assumptions the developing world passes the industrialized world in total GDP around the year 2020.[1] Total world GDP is projected to grow nine times between the years 2000 and 2100. Combining the GDP growth with the projected decline in energy intensity gives a fourfold increase in primary energy demand.

Market exchange and PPP

GDP numbers can be given using market exchange rates (what you get from a bank when changing one currency into another), or in terms of purchasing power parity (PPP; how much it costs to buy a defined basket of goods in different countries). For example, in China a few years ago a bank would give you eight of their yuan for each US dollar. That was the market exchange rate. If you bought a basket of goods in China and compared its cost to that of the same basket in the United States, you would have found that for each dollar spent in the United States you would need fewer than 8 yuan in China. That rate would be the PPP rate, and in this case it would be said that the Chinese yuan would be undervalued in the market. If the basket cost only 4 yuan, the PPP rate would be 4 to 1, not 8 to 1. Which rate you use affects both the apparent size of the economy and the energy intensity. However, it affects them in opposite directions and as long as you use the same system for both parts of the energy equation there is no problem. However, which you use can make a large difference when comparing standards of living.

The amount of greenhouse gases going into the atmosphere will go up proportionally to energy use if we continue business as usual with the same mix of fuels as is used now. World GDP has recently been going up rapidly. The larger developing countries are increasing their energy consumption and greenhouse gas emissions even faster than had been expected only a decade or so ago. The leading examples are China, India, Brazil, and Indonesia with others not far behind. Their rates of economic growth are much larger than those of the industrialized world and that is what drives their increased energy demand.

---

[1] The growth rates are after inflation and the evaluation is in terms of GDP at purchasing power parity (PPP).

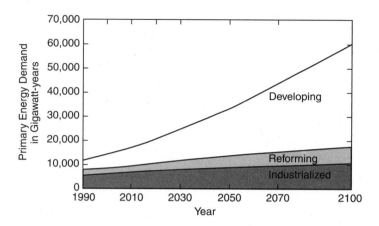

Fig. 6.2 IIASA projection of future energy-demand scenario A1 (high growth). IIASA projections show that energy demand in the twenty-first century is dominated by the growth of the developing nations. (*Source*: International Institute of Applied Systems Analysis and World Energy Council Global Energy Perspectives long-range projection; choose different assumptions here: http://www.iiasa.ac.at /cgi-bin/ecs/book_dyn/bookcnt.py)

Figure 6.2 is the IIASA projection of the growth of total primary energy supply (TPES) to the year 2100. Primary energy demand doubles by the year 2050 and nearly doubles again by the year 2100.[2] Most of the growth occurs in the developing countries. Around the year 2020 the developing world will overtake the industrialized world in total energy consumption. China overtook the United States as the largest emitter of greenhouse gases in 2008. In the business-as-usual scenario, emissions go up by the same amount as energy use increases, and CO_2e emissions in 2100 would be about 190 gigatonnes, which scares me as it should you.

6.3 EMISSION TARGETS

Global warming sneaked up on the world because of greenhouse gas emissions in the last century whose effects we did not begin

---

[2] Primary energy is the total energy content of the fuel used directly and to make other forms of energy. For example, a typical electric power plant is about 35% to 40% efficient. The primary energy used to make the electricity is two and a half to three times larger than the energy content of the electricity. While using electricity emits no greenhouse gas, making electricity certainly does. Similarly, gasoline refined from oil has only about 90% of the energy content of the primary oil, and in determining emissions the 10% lost in making liquid fuels has to be counted too.

to understand until recently. We still have much to learn, but have learned enough to realize that the nations of our world have to begin some sort of collective action to bring global warming under control. The first attempt, the Kyoto Protocol of 1997, set what was to be a binding commitment by the richer countries that signed it to reduce their emissions to roughly 5% below their 1990 emissions by 2012, the expiration date of the Protocol. Meeting the commitment requires that the average emissions during the 5 years from 2008 to 2012 be at or below the target. Most of the signatories will fail to make their agreed goal, but there will be benefits in showing what kinds of things do not work. I will come back to Kyoto and the reasons for its failure and how we might do better in the later sections on policy.

It is possible for the scientists to set an emissions goal that stabilizes the climate at some new and higher temperature relative to the preindustrial average temperature. As we saw in Chapter 5 there is still much uncertainty in how high the temperature will go for any given increase in the level of greenhouse gases in our atmosphere. Figure 5.2 showed that it will take 20 to 30 years to learn which end of the range of predictions we are heading for. We should not wait but instead plan to start with the average prediction of the temperature increase and adjust our program over the next 20 to 30 years as we learn more and reduce the uncertainties in the predictions.

I would try for stabilization at no more than double the preindustrial level which would be about 550 ppm of CO_2e compared with the 270 ppm of the eighteenth century. That gives a central value prediction of a temperature rise of 3 °C or about 5 °F. Many, particularly in Europe, claim that anything above a temperature rise of 2 °C or about 4 °F, which would correspond to 450 ppm of CO_2e, is dangerous. However, there is no sharp threshold to a danger zone and we are already at nearly 400 ppm CO_2. I don't think we can hold things much below 550 ppm by the year 2100. As technology improves, we may be able to do better. Figure 6.3 shows what has to be done to meet different goals.

I have a second reason for wanting to limit greenhouse gases to 550 ppm. The climate record from both the Greenland and Antarctic ice cores has examples of sudden temperature changes of many degrees. The temperature stays high for a while and then goes back down. Some of these instabilities last for centuries. We do not understand them and should be careful in our choice of allowed

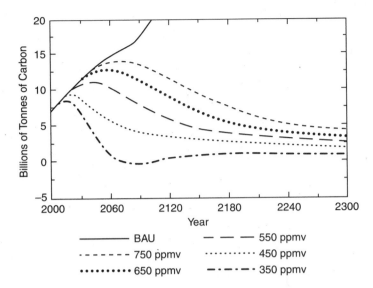

Fig. 6.3   Stabilization trajectories. Carbon emissions allowed if the atmosphere is to be stabilized at one of several new levels. Also shown is the BAU expectation. (Courtesy of Professor John Holdren, Harvard University)

greenhouse gas concentration. My choice of a limit of doubling eighteenth-century concentrations is a personal guess at how high we can go without greatly increasing the risk of a sudden large change, and I confess that there is no calculation that tells me it is the correct limit.[3]

Figure 6.3 shows examples of what are called stabilization trajectories. Also shown is the business-as-usual trajectory. (Note that this chart has emissions in terms of billions of tonnes of carbon; to get it in CO_2 terms, multiply the carbon emissions by 3.7.) A stabilization trajectory shows the allowed world greenhouse gas emissions versus time that will stabilize CO_2 at various levels. The reduction in emissions required to stabilize at 550 ppm can obviously be achieved more easily than those for 450 ppm. A target of 550 ppm also gives more time to bring the developing nations into the program as needs to happen if any of the stabilization scenarios is to achieve its goal (more about how this might be done in Chapter 17).

[3] The US Climate Change Science Program has recently released a report on abrupt climate change. It can be found at http://www.climatescience.gov/Library/sap /sap3–4/default.php

In any world program the industrialized nations will have to take the lead and reduce emissions faster than the developing world. Think of the richer countries running below the trajectory corresponding to the chosen goal, while the poorer counties run above it for a while. There are many more people in the poor countries today than in the rich ones, and as the poor ones improve their standard of living they will have to fully participate in whatever is agreed as the world goal. As noted earlier, China and the United States are the world's largest emitters, each contributing about 25% of the total (China's emissions today are slightly larger than those of the United States). However, China's per capita GDP (PPP) is about one-sixth of that of the United States. The policy problem is not only determining the goal, but how the reductions in emissions are shared among nations.

The shapes of these trajectories are also an issue. For any of the goals it is possible to let emissions go up for longer than shown and then bring them down more rapidly later, or bring them down earlier than the peak shown and more slowly later. The sooner we start the less chance that increasing temperature will cause serious damage to the world's ecosystem. However, the sooner we start the more the burden falls on today's economies and people. The economists have been arguing over how to do a massive switch to a lower-emissions world in the most economically efficient way. That is the subject of the next chapter, and I find that there seems to be more personal opinion than mathematics in the arguments of the opposing economic camps.

Technical Note 6.1:   Carbon dioxide equivalents

Different greenhouse gases have stronger or sometimes weaker effects on global warming than CO_2. The modelers work in terms of CO_2 equivalents or CO_2e. For example, CH_4 (natural gas or methane) is 21 times stronger as a greenhouse gas than CO_2. If I emitted 1000 tonnes of CO_2 and 10 tonnes of natural gas, the CO_2e of the combination is the same as adding the emission of 210 tonnes of CO_2 (the CO_2 equivalent of 10 tonnes of CH_4) to the original 1000 tonnes of CO_2. There are other gases that are even stronger than natural gas. Carbon tetrafluoride, used in microelectronics fabrication, is 6000 times as strong a greenhouse gas as CO_2. Fortunately, we emit very little of it. Table 6.1 shows

Table 6.1 *Global emissions in 2004 of the main greenhouse gases in gigatonnes of CO$_2$ equivalent, removal time, and percentage of each in the total*

Gas	CO$_2$ equivalent (Gt)	Removal time (years)	Percentage of total
CO$_2$ (fossil fuel)	27.7	>100	57
CO$_2$ (other industrial)	1.4	>100	3
CO$_2$ (agriculture)	8.5	>100	17
CH$_4$ (agriculture and industry)	7.0	10	14
N$_2$O (agriculture)	3.9	100	8
Fluorine gases	0.5	1000	1
Total	**49**		**100**

data from the 2007 IPCC Summary for Policymakers. It gives the annual emission of the main greenhouse gases in gigatonnes (billions of metric tonnes) of CO$_2$ equivalents (CO$_2$e).

7

How fast to move: a physicist's look at the economists

Former US President Harry Truman once said that he wished the government had more one-handed economists because his economists were always telling him on the one hand this, on the other hand that. Today, we do have many one-handed economists writing on the economics of taking action now to limit climate change. Unfortunately they seem to fall into camps with different hands. I will call the two camps after the two people who best represent them. One I call the Nordhaus camp after Professor William Nordhaus, the Sterling Professor of Economics at Yale University, and the creator of the Dynamic Integrated model of Climate and the Economy (DICE model) that is used by many to estimate the economic effects of climate change. The other I call the Stern camp after Sir Nicholas Stern, Head of the UK Government Economics Service, who led the effort to produce the influential 2006 British analysis of climate change impacts called the Stern Report (he is now Lord Stern of Brentford and is at the London School of Economics).

The issue is how much the world should be spending now to reduce the emissions that will cause large climate changes in the future. If we could assign a monetary value to future harm, and we used some reasonable discount rate (defined below), we could in principle figure out how much to invest now and in the future to reduce the harm. The Nordhaus camp says that we should be spending a reasonable amount now, but that there is no need to panic. The Stern camp in contrast says that the consequences are so severe as to constitute a global emergency and that drastic action is called for immediately. There is no one among the leaders of the economics profession who says we should do nothing now.

The surprise to me is that the two groups use economic models that are mathematically very sophisticated, seem on the surface to be

different, but are essentially the same. The difference in their conclusions comes down to one number, what the economists call the social
discount rate. Plug Stern's social discount rate into the Nordhaus
model and you will get the Stern result. Plug the Nordhaus value into
the Stern model and you will get the Nordhaus result. I conclude that
the difference is a matter of opinion, subjective rather than objective,
and, further, the more I looked at the analyses the more I came to
wonder if the notion of discount rates makes any sense when looking
hundreds of years into the future.

A discount rate is used to determine the present value of some
future thing. Look first at the simplest application. Suppose I want to
have one dollar (or euro or yen) 20 years from now. If I put money in a
bank that pays compound interest at a rate of 5% per year, I need to put
in only 38 cents today to have my account contain one dollar 20 years
from now. In economist terms, the discount rate is 5% and the present
value of that dollar 20 years in the future is 38 cents.[1] The present value
depends on the discount rate, which in this case is equal to the interest
rate. If the bank pays only 1% interest, I need to put in a larger amount,
82 cents, to make up for the lower interest rate. In this case, the discount
rate is 1% and the present value is 82 cents. If I can find a bank that will
pay 10% interest, the discount rate would be 10% and the present value
15 cents. Since the climate-change problem is one that requires centuries
to get under control, the time periods are longer and the present values
can be astonishingly small. Table 7.1 gives some examples.

The problem that we confront in climate change is that as the temperature goes up, we expect that there will be harm to the global ecosystem, and hence to the global economy. If we take the present value of
the harm and invest that much to avoid it, we economically break even.
Two problems are immediately apparent: how we determine the monetary value of the harm and how we determine the discount rate. This is
what the arguments among the economists are all about.

Let's look first at the discount rate. It is central to the economic
argument because as Table 7.1 shows, when you are looking ahead
for hundreds of years the present values change enormously as the
discount rate changes. In the economic argument, Table 7.1 is an over-
simplification. The economists add together two rates to get the total
discount rate. One of them is what might be called a wealth factor.

---

[1] Strictly speaking, the economists would say I am talking about the productivity
of capital, but using interest on a bank account is a more familiar way to talk
about the issues.

Table 7.1 *Present values in dollars of $1 for various discount rates and time periods*

Time period (years)	Discount rate			
	1%	3%	5%	10%
20	0.82	0.55	0.38	0.15
50	0.61	0.23	0.09	0.01
100	0.37	0.05	0.01	0.00001
200	0.14	0.002	0.00005	0.000000005

How much richer are the people in the future going to be? The US Bureau of Economic Analysis indicates that from 1980 to 2007, for example, the economy grew after inflation at an average rate of 3% per year, while the population grew by 0.9% per year. Take your choice on the wealth effect: 3% per year for the economy (120% richer for the total economy) or 2.1% for per capita income (75% richer per person [11]).

The other part is called the social discount rate and is a much fuzzier concept. Some call it a measure of impatience in that people value a dollar that they can spend now more than a dollar that they will spend later, which is in principle what a discount rate does. Other economists call it a measure of intergenerational equity; a value of zero says that later generations are as important to us as our own, while a large value says they should fend for themselves. In any event, Nordhaus says Stern uses a social discount rate that is much too small, 0.1%. This choice gives Stern's analysis a total discount rate that is much smaller than the historic rate. Stern says that a low social discount rate is the only ethical thing to do, and Nordhaus' value of 4% is much too large. Intergenerational equity in Stern's view must be a centerpiece of how the world handles problems today that will affect the welfare of the people of the future.

In a physicist's view (mine), I wonder if they are both trying to quantify the unquantifiable.

- Social discount rates are slippery things. I saved a lot for the benefit of my children and save a lesser amount for the benefit of my grandchildren; my children have the primary responsibility for their children. The social discount rate should be time-dependent and in any event is entirely a value judgment. It is not mathematically determined though you can use mathematics to determine its effects after you guess what is should be.

- I wonder if the notion of a discount rate makes sense over long time periods. If you are a business looking at the value of a new factory, or a government trying to decide between fixing roads and cutting taxes, it does make sense to use a discount rate related to the productivity of capital over a foreseeable future, probably about 10 to 20 years. However, over hundreds of years nothing is worth anything today with any reasonable discount rate.

- Stern sets a target for the maximum level of greenhouse gases we should allow. His target is an upper limit of doubling of the preindustrial level (which is what I would have recommended had he asked me). Nordhaus uses about the same thing. As I said earlier, I believe that above that level there is a real risk of climate instabilities that we do not now understand, but there is also a risk of such instabilities at a lower level. Nordhaus seems not to include a risk premium for a sudden instability (like an insurance policy) into his analysis, though it may be there and I just haven't found it. Stern's crisis analysis asks us to do the maximum possible mitigation now, so there is no more that can be done for an instability risk.

- Nordhaus does not seem to factor into his analysis how long it takes to fix the climate-change problem once things get bad. The removal time of the greenhouse gases is a few hundred to a thousand years. Stern calculates the total damage of inaction over a long period of time to justify the large savings coming from a large mitigation effort. Nordhaus criticizes Stern for calculating damage over a long period, but to me any economic analysis should look at consequences of inaction over the time it takes to fix the problem once it occurs.

The most important thing that the economists are trying to tell us is that many things that have long been thought of as free are not free. We have dumped greenhouse gases into the atmosphere and failed to understand that the atmosphere cannot digest it all without changing the economy as well as the ecology of the world. It is only in the past 20 years that we have begun to understand the economic as well as environmental damage that these emissions are beginning to cause and that the damage will grow over time. The economists urge that we make the costs of emissions part of the price of the good that produces them so that there will be a strong motivation to find better and cheaper ways to produce the product without the emissions.

In economist's terms, an externality is a cost borne by the society for something done by the producer of a product. Dumping greenhouse gases into the atmosphere at no charge is such an externality. Internalizing that externality makes the producer somehow include the cost in the price charged. If a coal-fired power plant emits lots of greenhouse gases, you should move in a reasonable time to add an emissions fee to the price of electricity from coal, which will encourage the replacement of coal with a less polluting power system, a windmill for example with no emissions.

Nordhaus and Stern both start with a price to be charged for emissions. Their initial prices are not far apart, but Stern ramps his up far faster than Nordhaus. When the reader thinks about who is nearer the truth, remember that the difference between them comes down to what we owe to future generations.

8

Energy, emissions, and action

8.1 SETTING THE STAGE

This chapter moves our discussion to how to reduce the effect of the energy we use on our environment. The amount of energy we use is so large that it is hard to get a feel for its size. I start with comparing the total primary energy supply (TPES) to natural phenomena that we could possibly use to supply the world's energy needs. The TPES from all sources amounts to a yearly average power of 14 terawatts (a terawatt is one billion kilowatts), a number that is too big to mean much to most people. It is the energy used to light all the world's light bulbs; run all the world's cars, trucks, buses, trains, airplanes, and ships; produce all the steel, cement, aluminum, and other metals; run our farms; produce all our computers; and everything else that we make or use.

In my time as a working physicist I did experiments involving subnuclear processes and processes that were related to the scale of our cosmos; from a billionth of a billionth of a meter to 14 billion light years. Those numbers mean something to me mathematically, but are not easy to visualize. So it is with the TPES. It is hard to understand what 25 trillion barrels of oil per year really is (it would cover the entire United States with oil one foot deep), or what many billion tons of coal is (six billion tons would give every man, woman, and child on Earth 2000 pounds of it), or what trillions of cubic meters of natural gas is (6 trillion cubic meters of gas would give each person 100 000 party balloons full of gas). Table 8.1 below is a comparison of what we use now to all the world's natural phenomena that can be used to generate energy. The table is for today, but the projection for increased energy demand is that the TPES demand will increase by four times by the end of this century while, of course, the natural sources remain

Table 8.1 *Natural energy sources compared to the total primary energy supply of 2004*

Item	Amount relative to TPES
TPES	1
Solar input	8000
All the world's winds	60
All the ocean's waves	4
All the Earth's tides	0.25
Geothermal world potential	2.3
All the world's photosynthesis	6.5
All the world's rivers	0.5

the same. As our demand goes up, the natural sources available seem to get smaller relative to what we will need.

All of the natural energy is not really accessible. Oceans cover 70% of the world. We cannot absorb all of the wind without disastrous climate problems. If we used all the photosynthesis, there would be no flowers to admire or food to eat. There is no sunshine at night. The wind does not blow at any one place all the time. However, the potential resource is large, and the transition from mining the fossil fuels laid down over millions of years to basing our economies on what might be continually available is not a hopeless dream. It will not be easy and it does not have to be done in a very few years. Indeed, I do not believe it can be done in as few as 10 years, in spite of what some well-known advocates say. What the possibilities and limitations are is the subject of this section of the book. We would also do well to remember that while the energy sector is the largest part of the climate-change problem, it is not the only part. The agricultural sector will be the hardest to control, and too little effort has gone into understanding it. Agriculture will need to be included in emissions reduction eventually, but we can begin with the energy sector which we do understand and where the largest early reductions in emissions can be made.

8.2 SOURCES OF EMISSIONS

Chapter 6 showed the main sources of the greenhouse gases that are causing climate change. About 70% of our emissions come from fossil-fuel use and some industrial processes, while about 30% comes from agriculture and land-use changes. Fossil-fuel use is what I focus on here. We can do a lot to reduce emissions from our energy supply;

it will be relatively easy at the start because there are large gains in efficiency to be made, and become progressively harder as the CO_2 emissions are wrung out of the system.

It is important not to mix agendas as those I called the ultra-greens seem to do. If carbon can be captured and put away securely, do it. If you substitute a modern gas-fired electricity generating plant for an old coal-fired one and eliminate two-thirds of the previous emissions, do it. You do not have to replace everything with solar power or windmills.

The distinction between primary energy and the energy directly used by us is important. Emissions come mainly from primary energy while what we use is a mix of primary and secondary energy. For example, electricity is a secondary form of energy and has to be made from something else – there are no electricity mines. Electricity is made at power plants that use some primary fuel, coal or natural gas, or nuclear reactors, or wind, or water, etc. Electricity, then, is not primary energy. If you drove an electric car you would use no gas-oline and emit no greenhouse gas directly, but producing the electricity you used to charge your batteries to run the car would produce greenhouse gases, and how much depends on the mix of fuels used to generate the electricity. On the other hand, if you use natural gas to heat your home, you are directly using one of the primary energy sources. To reduce our greenhouse gas emissions we have to reduce the emissions from the primary fuel supply. So, if someone tells you of the marvels of hydrogen as a motor fuel because using it emits no greenhouse gases, ask how the hydrogen was made and what those emissions were.

The primary energy supply consists of six main fuels: oil, coal, natural gas, nuclear energy, large hydroelectric energy plants, and combustibles (mostly forest and agricultural waste material). What are called "Other" in Table 8.2 below include wind, solar, geothermal, small-scale hydropower, and biofuels which together only amount to about 1% of TPES. The breakdown by fuel is important because their use in the economy differs and the roads to decarbonization of the various sectors of the economy also differ. For example, in most of the world, oil dominates the transportation sector, coal dominates electricity generation, and gas is the largest source of heat for buildings.

Table 8.2 below gives the breakdown of the components of the primary energy supply worldwide according to the International Energy Agency's most recent analysis [11]. The data are from the year 2005 and the projection under business as usual is that the breakdown by fuel will be about the same in the year 2030, though the total usage will be

Table 8.2 *Percentage of total primary energy supply (TPES) and world CO_2 emissions by fuel*

Energy source	Percentage of TPES	Percentage of world CO_2 emissions
Oil	34	40
Coal	26	40
Natural gas	21	20
Nuclear power	6	0
Hydroelectric	2	0
Combustibles	10	0
Other	1	0

up by about 50%. This is consistent with the older International Institute of Applied Systems Analysis (IIASA) projection that I used earlier.

Of the three fossil fuels, on an equal-energy-content basis coal is the largest greenhouse gas emitter, oil is next and natural gas is last. Even though oil represents a larger piece of the TPES, oil and coal are each responsible for about the same greenhouse gas emissions.

Energy and greenhouse emission from fossil fuels

Coal's energy comes from turning carbon into carbon dioxide, so all of the energy from coal comes with greenhouse gas emissions. Oil is mainly a complex molecule with roughly two hydrogen atoms for each carbon atom, and its energy comes from turning one carbon atom into carbon dioxide and two hydrogen atoms into water (H_2O). Only that part of the energy coming from turning carbon into CO_2 produces greenhouse gas, and so oil produces about 75% of the CO_2 emission of coal for the same primary energy. Natural gas has four hydrogen atoms per carbon atom and its energy comes from turning the carbon into CO_2 and the hydrogen into two molecules of water. Of the primary energy produced by burning gas, only about half comes with CO_2 emissions and so natural gas emits 50% of the CO_2 of coal for the same primary energy.

Combustibles are largely the plant material gathered by the world's poor to supply heat; they are not the biofuels so talked about today. Since plants get the carbon for their growth from the CO_2 in the atmosphere and release it on burning, they do not give any net

Table 8.3 *US energy flow in 2006*

Primary energy	$\rightarrow$	End-use sector
Coal (40%)		Residential buildings (21%)
Gas (22%)		Commercial buildings (18%)
Oil (23%)		Industry (32%)
Nuclear (8%)		Transportation (29%)
Renewables (7%)		

Source: EIA *Annual Energy Review* (2007) [14]

increase in greenhouse gas as long as they are grown without fertilizer and other modern agricultural technology. They are part of the Photosynthesis component listed in Table 8.1.

The category "Other" includes all of what are called Renewables: biofuels (which do have greenhouse gas emissions), wind, solar, geothermal, etc. They are a negligible component of the energy mix today, but are expected to grow. Their problems and promises are the subject of later chapters.

The US breakdown for total primary energy supply by fuel is not very different from the world breakdown shown in Table 8.2. Table 8.3 shows a different kind of breakdown, where the primary energy goes by sector in the economy. The primary sources supply the end-use sectors in different ways. For example, most of the oil is used for transportation while most of the coal goes into producing electricity which goes to all of the sectors. All of the nuclear and most of the renewables go to electricity production, while gas is split between heating and electricity.[1] Chapter 11 shows the breakdown by fuel and by sector.

US greenhouse gas emissions from fossil fuels in 2007 totaled about 6300 metric tonnes of CO_2e out of a total of about 7000 metric tonnes, the difference mainly due to agricultural emissions [12]. Decreasing emissions requires an attack on the emissions from the energy sector.

8.3 REDUCING EMISSIONS

The only way to reduce global greenhouse gas emissions while maintaining the same economic output is to change the mix of fuels we use, to use energy more efficiently, or a mixture of both. To see why

[1] For a more detailed breakdown see the DOE EIA *Emissions of Greenhouse Gases Report* (2007), http://www.eia.doe.gov/oiaf/1605/ggrpt/index.html

this is so, I want to expand on a way of looking at things that I introduced in Chapter 6. There I discussed a way to get at energy demand in the seemingly roundabout way of considering GDP (the size of an economy), and energy intensity (the amount of energy used to produce a unit of GDP). The amount of energy used by a country, a region, or the world could be determined from

$$\text{Energy} = \text{GDP} \times (\text{Energy/GDP}).$$

This formulation is useful for energy because we know (or at least think we know) how to predict future GDP from projections of economic growth, and how to predict future energy intensity from historic trends.

I want to do the same sort of thing with greenhouse gas emissions. The way to look at emissions is as a product of four things: population multiplied by per capita income (GDP divided by population) multiplied by our old friend energy intensity (energy divided by GDP) multiplied by a new term called emission intensity (emissions divided by energy). Just like the equation in Chapter 6 this seemingly roundabout approach lets us get at emissions using things that are easier to estimate. This is the equation below, and if emissions are to go down, one or more of the components of the emissions equation have to go down.

$$\begin{aligned}
\text{Emissions} = {}& \text{Population} \\
& \times (\text{GDP/Population}) \\
& \times (\text{Energy/GDP}) \\
& \times (\text{Emissions/Energy})
\end{aligned}$$

World population is going to go up for at least the next 50 years according to the UN's projections. There is no help there. The next piece, GDP/Population, is per capita income. The poor of the world are not willing to stay poor, and the rich of the world are not willing to become poorer for the sake of their currently poor brothers and sisters. Per capita income is going to go up for a while at least. There is no help for the foreseeable future in that term either. All emissions reductions have to come from the next two parts of the equation, energy intensity and emissions intensity.

The third piece, energy intensity (Energy/GDP) can change dramatically if the efficiency of using energy is improved. Efficiency improvements are not necessarily about lifestyle changes but how to do the same with less energy. Lifestyle changes are advocated by some as well, and if they are significant can also reduce emissions by a large amount. However, that is a societal decision with large political implications and is not a topic I intend to discuss. There is a huge amount of room for efficiency improvements, and Chapter 11 will look at the possibilities.

The fourth piece of the equation, emissions intensity (Emissions/ Energy), is the emissions from the fuels used to power the world's economies. Changing from a fuel with high emissions to one with low emissions obviously reduces emissions, and this is what is behind the move to carbon-free or low-carbon energy sources. We can start this move in several ways: substituting natural gas for coal (the reduced carbon path); or substituting sources that have no carbon like wind or nuclear power for fossil fuels (the carbon-free energy path); or capturing and putting away the emissions from fossil fuels (called carbon capture and sequestration or CCS), which is the hope of the coal industry; or doing the same thing with less energy (the increased efficiency path). Carbon-free energy is already available on a large scale from hydroelectric and nuclear power plants (life-cycle emissions including those from plant construction are shown in Chapter 10). Wind and solar power, though still small, are beginning to have significant market penetration. Substitute a solar-energy electricity power plant for a coal-fired plant and emissions go down. Unfortunately, things are never as simple as we would like. The sun doesn't shine at night and the wind doesn't blow all the time. As of now we have no good way to store electricity made from these intermittent sources, so they cannot do the job alone. What we can and cannot do in the near term is discussed in later chapters on all the energy technologies.

8.4 NO SILVER BULLETS

In folklore the silver bullet can slay a monster, vampire, or werewolf. The consequences of a major temperature increase from greenhouse gas emissions certainly qualify as an evil monster because of the effects on our civilization. Sea level will rise, tropical diseases will move north in the Northern Hemisphere and south in the Southern, crop growing seasons will change, rainfall patterns will shift, major storms will grow stronger, and the list goes on. It would be wonderful if there were a single silver bullet that could slay the climate change monster, but there isn't.

We need to recognize that solving the problem will be hard and the solution will have to start with approaches on many fronts. A useful way to think about the problems is in terms of what two Princeton

University scientists called "stabilization wedges" in an important paper published in 2004 [13]. In the business-as-usual scenario, emissions are going to continue to go up as the world uses more energy. What we need to do to stabilize greenhouse gases in the atmosphere at some desired level is to reduce what we emit. We can do that by a collection of approaches – the stabilization wedges – such as increased miles per gallon for cars which reduces emissions from gasoline, or substituting emission-free wind turbines or nuclear power for coal-fired power plants, or improving insulation in buildings so they use less energy to cool in summer and heat in winter. Wedges can be anything that cuts the emissions of greenhouse gases. Some of the wedges will be wide while others may be narrow, but collectively they get us to where we want to go as long as there are enough of them. Figure 8.1 shows the idea.

The job of the scientists, technologists, and industrialists is to develop the things that can make up stabilization wedges. We know what many of them can be today, but there can be many more in the future with the proper support for development and incentives for deployment. It is the job of governments to provide that support and to create the collection of incentives and sanctions that will help assure the adoption of the techniques that go into the wedges.

The wedges always start small and grow over time. No major change in our energy system can have an instantaneous large effect. If, for example, I want to introduce natural-gas-fired electrical power plants instead of coal-fired ones because I can reduce emissions with gas as a fuel instead of coal, I have to build the plants one at a time and the benefit builds up over time. If I have a new technology that makes automobiles go twice as far on a gallon of gasoline I have to start producing the cars, get people to accept them, and over time replace the old-style auto fleet. That process takes 10 to 20 years. The next several chapters will review our options including:

- Fossil fuels – how much there is, how long they will last, how they might be better used;
- Efficiency in transportation and buildings – enormous gains can be made by doing the same jobs with less energy thus reducing greenhouse gas emissions while also reducing what we spend on energy;
- Carbon-free and reduced carbon energy – benefits and limitations of nuclear, wind, solar, geothermal, and biofuels.

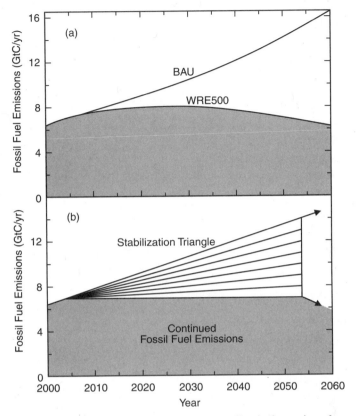

Fig. 8.1 Stabilization wedges. (a) The upper line is the projected emission with business as usual while the lower line defines the maximum emission allowed to stabilize the concentration of CO_2 in the atmosphere. (b) The space between the projected and allowed emissions is filled with non-emitting stabilization wedges. (*Source*: S. Pacala and R. Socolow [13], reprinted with permission from the American Association for the Advancement of Science)

8.5 WINNERS AND LOSERS

Normally in an evaluation of the potential for greenhouse gas reductions from various energy options, I would go through all the options and then give a summary. I will do that in Chapter 15, but will also give my score card now so the reader knows where I am going and can pay closer attention to topics where there is some disagreement.

Winners:

- **Efficiency** in all sectors (if you don't use it, it doesn't emit)
- **Coal** (with carbon capture and storage)
- **Hydroelectric**
- **Geothermal** (near-surface systems)
- **Nuclear**
- **Natural gas** (as a replacement for coal)
- **Solar heat and hot water**
- **Sugarcane ethanol**
- **Solar photovoltaic** (for off-grid applications only)
- **Advanced batteries (**for plug-in hybrid or all-electric vehicles)

Losers:

- **Coal (**without carbon capture and storage)
- **Oil** for transportation (replaced with electric drive)
- **Corn ethanol**
- **Hydrogen** for transportation

Maybes:

- **Enhanced geothermal** (deep mining for heat)
- **Solar thermal electric** (needs cost reduction)
- **Solar photovoltaic** (large subsidies needed, so only for the rich now)
- **Advanced biofuels**
- **Ocean systems**
- **New technologies not yet invented** (remember it is hard to predict the future because it hasn't happened yet)

9

Fossil fuels – how much is there?

The world economy runs mainly on fossil fuels – coal, oil, and natural gas – and as shown in the previous chapter, they are the main sources of greenhouse gas emissions outside the agricultural sector. They were made in geological processes that turned plants grown hundreds of millions of years ago into the fossil fuels we use today. High temperature and high pressure can convert a prehistoric tree into a piece of coal, or under different conditions of temperature and pressure into oil or gas. What we are doing today is mining the fuels generated so long ago at a rate much faster than they can be replaced by the processes that produced them in the first place. This means that fossil fuels are going to run out eventually and the era of powering the world economy with them will come to an end. The question is not if, but when, so the movement away from fossil fuels that is required to deal with climate change will eventually have to happen anyway to deal with resource exhaustion. Think of this century as a transition period in a move away from the energy resources that have brought great economic benefits, but have turned out to bring an unexpected problem – global warming.

Some say the era of available and affordable fossil fuels is coming to an end very soon, but the data on reserves say this is not true. There is enough coal, oil, and gas to last for a good part of this century even under the business-as-usual scenario, and the rate at which exploration has added to proven reserves has exceeded the rate of consumption of those reserves for many years.[1] However, if the growth in demand continues at its present rate it is very likely that there will be

---

[1] A word of caution: government estimates tend to be on the optimistic side. I have tried to find conservative estimates, but be warned: there may be less than is indicated here for all except natural gas.

supply constraints in the second half of the century. All the fossil fuels will certainly get more expensive as the easy-to-access sources begin to be used up. Part of the increase in the price of oil seen in recent years is because meeting demand has required tapping resources like the Canadian tar sands, where production costs are much more expensive than the standard light oil produced by OPEC.

9.1 WORLD OIL RESERVES

World oil consumption in 2008 amounted to about 85 million barrels per day according to the Energy Information Administration (EIA) of the US Department of Energy; in energy terms, about 34% of TPES. Most of the oil goes to fuel the transportation sector (95% of the transport sector runs on oil). The rest goes into petrochemicals, heating, pesticides, some industrial processes, electricity production (a small percentage), cosmetics, chewing gum – you name it. Oil demand worldwide is projected by the International Energy Agency (IEA) to increase at the rate of 1.6% per year. If that rate of increase were to continue for the rest of the century we would be using more than 300 million barrels per day by the year 2100 and would be in trouble with supply, as I show below.

Some have said that there will be a peak in oil production in the next few decades followed by a slow decline, but this is only partly true. There will be a peak in the next few decades in oil with low extraction costs, but not in oil as a whole. The notion of a peak followed by an inexorable decline is based on a simple idea. The resource has a certain size; demand keeps rising; as more and more of the resource is used up it becomes harder to keep up production, and production has to begin to fall. In the oil business this is known as Hubbert's peak after Dr. M. K. Hubbert. He was a geologist specializing in oil and predicted in 1956 that US domestic oil production would reach a peak in the early 1970s and fall thereafter. His prediction was based on the speed with which demand for oil was increasing and the (decreasing) rate at which new reserves were being discovered. Domestic production in the United States did peak in 1971, and ever since the peak notion has attracted much attention and resulted in many wrong predictions. The problem is that the resource has to be known for the time of peak production to be determined, and the resource keeps on expanding as new kinds of oil are added to the world reserves that were not previously considered to be usable. What Hubbert really said was that there would be a peak in the production of the light, easily extracted oil we were producing then, and there he was correct. In the oil industry, technology has been developed that makes production possible from resources that were previously thought to be unusable, and this trend continues which builds up reserves through technology advances as well as through the more normal discovery.

The oil industry breaks supply down into several types. What we pour into our gas tanks or into our engines is a product of an oil refinery where the raw oil has been treated to change its characteristics. What treatment is required and how expensive it is in both energy and money depends on the type of oil. The first breakdown made in the industry is into conventional and unconventional oil. Conventional oil includes what is called light sweet crude like that from the North Sea or Texas, sour crude like much of the oil from the Middle East, and some of the heavy oils. Light sweet crude requires the least treatment. Sour crude has high sulfur content and requires more refining. The boundary between conventional heavy oil and unconventional heavy oil is a fuzzy one that is loosely defined by what has to be done to get it to flow in a pipe. If it flows by itself it is conventional. If it has to be heated or treated somehow it is unconventional. Figure 9.1 gives the IEA estimate of reserves as of the year 2005 [15].

The figure shows the size of the reserves and the cost of production from a particular type of reserve. Since the costs are given in 2004 dollars, the numbers have to be adjusted for inflation and for the decline in the dollar relative to other currencies to get them to current levels. They are probably about 20% higher in 2008 dollars. Everything up to and including enhanced oil recovery (EOR) is conventional oil. To put the reserves into perspective, at present consumption rates the world uses roughly 30 billion barrels per year now, has already consumed about a trillion barrels, and with the 1.6% per year rate of increase will consume the next trillion barrels in less than 30 years. It is startling to realize that the next 30 years are expected to use as much oil as the entire world production up to now.

Middle Eastern OPEC oil (OPEC ME) is the least costly to extract, and there is estimated to be more than another trillion barrels available. Other conventional oil is that coming from the rest of the world's reserves, and there is nearly another trillion barrels that are expected to be available. Deepwater and Super Deep are what is thought to be available offshore in waters much deeper that have been exploited to date (there is so little of it compared with demand that the recent chant "drill baby drill" as a solution to the huge oil imports problem in the United States is obviously silly). Arctic is what is thought to be

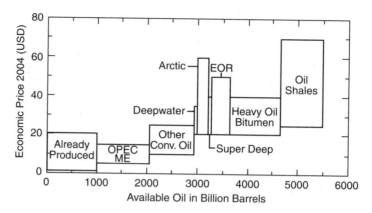

Fig. 9.1   Availability of oil from various sources and their prices. The range of prices and amounts available for various oil sources including estimates of technological progress in oil extraction. (*Source: Resources to Reserves, Oil and Gas Technologies for the Energy Markets of the Future* © OECD/IEA, 2005)

available in all the world's arctic regions. Average oil extraction from an oil field with today's technology is typically only about 35% of the total available. EOR defines the extra that would be available if plausible enhanced oil recovery technology improved the recovery rate.

Heavy oil, bitumen, and oil shale represent the unconventional resources. Their extraction and refining cost energy as well as money. The former Chief Scientist of the British oil company BP, Dr. Steven Koonin, told me that the Canadian tar sands require 15% to 30% of their energy content to extract the material and turn it into useful oil. The heavy oil and bitumen that we know about now are located mainly in Canada and Venezuela while oil shale is present in large amounts in the United States. Exploration of the world continues, and new resources are likely to be found. This is especially true for unconventional oil. Keep in mind that these resource estimates are uncertain. There may be more.

Four and a half trillion barrels yet to be extracted sounds like a lot of oil, but it will not even last to the end of this century if the projected yearly rate of increase in consumption is really as large as 1.6%. At that rate consumption doubles every 45 years, and the 4.5 trillion barrel reserve would run out in only 75 years. The cost of oil will go up as the world comes to depend on the harder-to-extract unconventional oil. An oil optimist who did not care much about emissions would say that it is possible that huge new reserves will be found and that the supply problem will vanish. A realist would say that is very unlikely and we had better start doing something about oil consumption. Oil will start to become much more expensive after 2030 when we have used the second trillion barrels, and it will eventually run out. Other methods of fueling transportation will be needed. I will discuss how and what in the chapter on efficiency.

9.2 WORLD GAS RESERVES

Worldwide natural gas consumption today amounts to about 2.95 trillion cubic meters (TCM) per year. In energy content this is equal to 21% of TPES. The IEA projection is that gas demand will increase by about 2.3% per year. The Middle East, particularly Qatar, and Russia have about 70% of the 180 TCM of proven reserves. The estimate of total conventional reserves is about 370 TCM, but exploration has not been as thorough as it has been for oil and there may be much more. If there is not, there will be insufficient natural gas to last until the end of the century.

Like oil reserves, gas reserves are divided into conventional and unconventional reserves. Unconventional in the gas case means anything different from that recoverable from the standard wells, and these reserves are estimated to be another 250 TCM. We know now that there is much more unconventional gas; the problem is getting at it. New technologies have made gas accessible from coal beds too deep to mine, and from gas trapped in the oil shale beds that are abundant in the United States. The shale beds are typically deep underground and not very permeable to gas movement. If old-style vertical wells were needed to extract the gas it would be far too costly because many wells would be needed over the entire area of the shale bed. The amazing technology developed in the oil industry of turning a drill from a single well from vertical to horizontal has made it practical to run an extraction pipe for long distances through a relatively thin layer and extract gas over a long range with one well as is sketched in Figure 9.2. Because of this new technology, US reserves have been increasing rapidly and those in the rest of the world will probably follow.

A much larger source has yet to be tapped at all. Trapped in the permafrost of the far north and in the ocean is gas in the form of methane hydrate. At low temperature and high pressure, natural gas mixed with water forms a kind of ice that traps the methane. At depths of a thousand feet (300 meters) in the Arctic permafrost and at depths greater than about 1500 feet (500 meters) in the oceans the

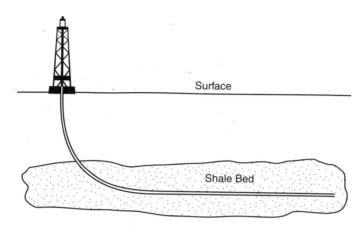

Fig. 9.2   Horizontal drilling to enhance gas recovery from shale beds. New drilling technology makes gas production economical from previously unusable sources.

conditions are right for the formation of this peculiar substance. The IEA estimate of recoverable gas from this source range from 1000 to 10 000 TCM, but the resource may be much larger.[2]

There is no commercial-scale gas production from the hydrates. Japan is probably the world leader in research and development (R&D) aimed at learning how to recover gas in this form. If it can be done, reserves will greatly expand. The US had a long-term R&D program on hydrates that was terminated in 1992 in favor of short-term, production-oriented R&D, a mistake that is typical of many on-again/off-again energy programs run by the United States. Methane hydrate recovery has to be done carefully because methane is a much stronger green-house gas than CO_2 and if the recovery process leads to significant releases of methane to the atmosphere, we will be in a lot more climate change trouble than we are now. How expensive it will be to tap this resource in an environmentally sustainable way is still unknown.

The required amount of gas to last the century can be estimated as it was for oil. Demand is growing by 2.3% per year and consumption today is 2.95 TCM per year. The amount required up to the year 2100 will be about 900 TCM. Unless reserves are much larger than estimated today or the hydrates turn out to be usable, we do not have enough gas to continue business as usual. Those who advocate a switch from

[2] Methane hydrate is one of a more general class of mixtures called clathrates. Even CO_2 can form a clathrate and this is the basis of the idea that perhaps CO_2 can be sequestered as a clathrate in the deep oceans.

Table 9.1 *Coal reserves and consumption*

Country	Known reserves 2003 (billions of tons)	Consumption 2007 (billions of tons)
United States	271	1.1
Russia	173	0.26
China	126	2.3
India	102	0.51
Australia	87	0.15
South Africa	54	0.20
Rest of world	188	2.7
World total	1001	7.2

Source: DOE Energy Information Agency

coal to gas because of the smaller greenhouse gas emission for the same energy output should recognize that this can only be a temporary measure. A switch to non-emitting fuels is required for gas just as for oil.

9.3 WORLD COAL

The coal story is not very different from the oil and gas stories. The US Energy Information Agency (EIA) estimates the growth in coal demand as 2.6% per year from now to 2015 and then 1.7% per year from 2015 to 2030. For simplicity I will use the 1.7% annual growth number for the entire century. Table 9.1 shows the reserves and current consumption for the world's largest coal producers and for the entire world.

Using the EIA 1.7% per year growth in demand starting now, the coal reserve will run out in the year 2080. However, the situation is really more complicated than a simple look at reserves and demand indicates. There are different types of coal containing different impurities and different amounts of carbon. The power plants that use these different types are designed to handle a specific type, and it is no easy matter to convert a plant designed for one type to use another. Perhaps there is much more coal; reserves are hard to estimate. Still, there should be no problem with supply in the first half of this century.

9.4 CONCLUSION

The world can continue its profligate use of fossil energy for the next 50 years. Beyond that there is only uncertainty. If the reserve estimates discussed above are anywhere near correct, supplies of fossil fuel will become harder to come by and more expensive as the more difficult unconventional reserves begin to be tapped on a larger scale. In addition, there is no substitute for some kind of carbon-based material in production of items ranging from petrochemical to plastics to chewing gum, and therefore some fossil fuel should be saved for uses other than energy.

All the fossil fuels have had large price increases in the past few years. While those of oil have been the most dramatic, coal and gas prices have gone up by two to three times. Prices have dropped recently (early 2009) with the slowdown of the world economy. The current recession will end, as have all the previous ones; when it does, economic growth will resume and with it demand for energy will grow again. Unless new fossil energy reserves are discovered, the recent price increases are small compared with what will come later. There is more reason to develop carbon-free energy sources than combating global warming.

I always use the term carbon-free energy sources rather than renewable energy sources which are more limited. The renewables are generally taken to include solar, wind, geothermal, hydroelectric systems (sometimes only small hydroelectric projects are included in renewables), and biofuels. Carbon-free would also include energy-efficiency programs, fossil fuel with greenhouse gas capture and storage (CCS), large-scale hydroelectric where appropriate, nuclear energy systems, and perhaps even nuclear fusion eventually. Efficiency reduces energy use and thereby emissions. CCS does not get the world away from the problem of the potential exhaustion of fossil fuels, but it does allow fossil fuels to be used for a longer period and give more time for the development of better and less costly fossil-fuel-free sources. Large hydroelectric dams are useful in some parts of the world. There are enough fuel reserves already known for nuclear reactors to last for thousands of years, and enough in the sea to last for tens of thousands of years if economic extraction systems can be developed. The focus for now should be on pushing the development of all systems that can give the world a safe and effective energy supply that is free of greenhouse gas emissions. In the next chapters I will go over these.

10

Electricity, emissions, and pricing carbon

10.1 THE ELECTRICITY SECTOR

Worldwide, the two largest sources of greenhouse gas emissions are electricity generation and transportation. Electricity generation is the topic of this chapter while transportation is part of the next.

Coal makes up by far the largest fraction of fuel used to produce electricity. The United States and China are the Saudi Arabias of coal, and coal with all its emissions problems is the fastest expanding fuel for electricity production. It is the lowest in cost because of its abundance and ease of extraction, and because power plants can be built relatively quickly. Without some sort of emissions charge or other limitation mechanism, coal will remain the lowest-cost fuel for a long time to come. Finding a substitute for coal or a way to reduce emissions from coal is critical to the world effort to reduce greenhouse gas production.

In the United States, coal and natural gas are used to generate 70% of electricity, and, according to the EIA, are responsible for producing nearly 40% of US greenhouse gas emissions. The percentages are not very different from those of other industrialized countries with the exception of France. France gets most of its electricity from greenhouse-gas-free nuclear power and has much lower emissions per unit GDP and very much lower emissions from the electricity sector. I will come back to this in the chapter on nuclear power.

There are only four ways to go about reducing the emissions from electricity generation:

- Emit less greenhouse gas by making electricity generation more efficient (less fuel for the same electrical output);
- Catch the greenhouse gases and store them away;
- Use electricity more efficiently (less demand means less generation which gives lower emissions);

- Substitute low- or zero-emission sources for what we use now (solar, wind, or nuclear power, for example).

I start with the efficiency of the existing electrical generating system which is not nearly as good as it could be, go on to what the charge for carbon emissions might be if the costs to society that come from emissions were to be included in the price of electricity generated from fossil fuels, and end with catching and storing the greenhouse gases, something I am skeptical about, but which has such potential that it is worth a try. End-use efficiency and carbon-free or low-carbon sources are discussed in later chapters.

The reason for the inefficiency in generation and the larger-than-needed emissions that go with it is a combination of low fuel costs for fossil-fueled generation plants, and ignorance of the consequences of greenhouse gas emissions until relatively recently. Most of the US electrical generating plants are old with an average age of 35 years. When they were built, fuel was cheap and global warming was a thing few scientists worried about, much less citizens or governments. As a result, the US electricity supply (and the world's) has come to depend more and more on coal, which has been the lowest-cost generator of electricity and still is today. Figure 10.1 from the EIA [14] shows the evolution of the US electricity supply from 1949 to 2007. Coal is king of electricity generation, and with its crown has come a large increase in emissions. It now supplies 50% of the electricity while gas and nuclear supply roughly 20% each, hydroelectric dams supply 7% and the renewables supply the rest. The same report [14] also shows the flow of energy into the generating system and the flow of electricity out. From it you can find that:

- Only 35% of the primary energy in the fuel gets transformed into electricity;
- Of that, some of the electricity is used inside the power plants, some is lost along the way in the distribution system, so only about 31% of the fuel energy reaches the consumers in the form of electricity;
- Gas is better than coal – 40% of the energy in gas is turned into electricity, but only 33% for coal.

The emissions in electrical generation depend on the fuel used in the power plant. I often get questions when talking about low- or zero-emission sources of power about what happens if I include the emissions in manufacturing the power plant as well as the emissions from its operation. The answer is in Figure 10.2 which shows the life-

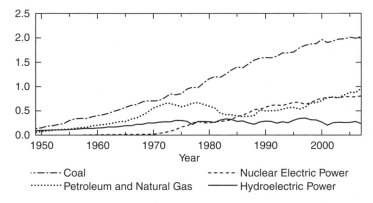

Fig. 10.1   Evolution of the electricity supply. Since 1950 electricity production has come to be dominated by coàl. (*Source*: EIA *Annual Energy Review 2007* [14] Fig. 44, http://tonto.eia.doe.gov/FTPROOT/multifuel/038407.pdf)

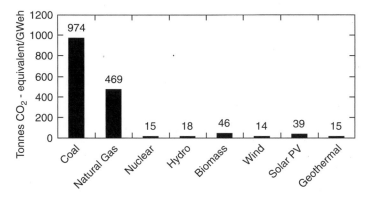

Fig. 10.2   Comparison of life-cycle emissions in metric tonnes of CO_2e per GW-hour for various modes of electricity production. (*Source of data*: [16], [17], [18])

cycle emission per gigawatt-hour of electrical output for coal, gas, and several forms of carbon-free energy (not totally carbon-free but nearly so) [16, 17, 18]. Life-cycle emissions take the total emissions coming from everything except burning the fuel (making the steel and concrete used in the plant, mining and transporting the fuel, maintaining, repairing and upgrading the plant, etc.), averaging them over the expected lifetime of the plant to get non-fuel emissions per hour and adding that to the direct emissions from operations. In this kind of analysis even wind power, which uses nothing but the natural wind to generate electricity, has some life-cycle emissions from

Table 10.1 *Comparative risks of different energy sources expressed as years of life lost per billion kilowatt-hours, based on an end-to-end analysis for Germany*

Coal	138
Oil	359
Natural gas	42
Nuclear	25
Photovoltaic	58
Wind	3

Source: Ref. [19]

manufacturing and maintaining the wind turbines. Replacing coal plants or gas plants with any of the carbon-free sources can make big reductions in greenhouse gas emissions even when every input is included. Even replacing coal plants with gas plants makes a big reduction in emissions.

It is also worth taking a look at the comparative risks of all the main sources of electricity. Table 10.1 summarizes estimates made by W. Krewitt and colleagues on the comparative risks of various energy systems [19]. Krewitt's analysis is for Germany, is based on European Union regulations for emission controls, and uses German government numbers for health effects. His comparisons are for the same amount of electricity generation from each source, one terawatt-hour (one gigawatt of electricity for 1000 hours), and are supposed to be end-to-end; that is, to include mining, transportation, fuel fabrication, plant construction, operation, and the effect of emissions on public health. Oil is the worst, but oil is used for only a small fraction of electricity generation. The fact that coal is the worst of the major fuels for electricity generation is no surprise. The biggest surprise for me was the large number for photovoltaic generation. Krewitt's analysis is based on the use of polycrystalline photocells which use many toxic gases in fabrication. I would expect the thin-film cells coming into use now to be less hazardous, but there are not enough of those deployed to make a good estimate as yet. The analysis does not deal with uncertainties as well as I would like and the numbers give an impression of precision that I think is unwarranted. Nonetheless, of the major fuels, coal is clearly the worst, wind the best, and nuclear seems somewhat better than photovoltaic or gas.

The efficiency of electricity generation is important in discussing emissions. The aging fleets of US coal and gas-fired power plants run at average efficiencies of about 33% and 40% respectively, while the most modern coal plants run at 48% and gas plants run at 60%. Though the United States prides itself on technology, it is a long way from the best in the world when it comes to efficiency of electricity generation [20]. Denmark is best with coal at 43% efficiency while Luxemburg and Turkey lead the way with gas at 55%. Honors for being the worst go to India for coal at 27% and Russia for gas at 33%.

If all of today's coal and gas plants in the United States were converted to even 50% efficient natural-gas plants, emissions from electricity generation would drop by nearly 1500 million tonnes per year or nearly 25% of total emissions. Gas plants are both more efficient than the best of coal plants and emit less greenhouse gas for the same amount of energy used[1] so wherever there is enough gas available, conversion from coal to gas should be encouraged.

The reasons for our continued reliance on coal are that there are a lot of plants already in existence, and, more importantly, coal is cheaper than gas as a fuel. Even though a coal plant with all its pollution-control equipment costs more than a natural-gas plant to build, the difference in fuel costs make coal a source of lower-cost electricity (according to the EIA, in 2007 gas costs more than three times coal for the same energy content). If the unconventional gas resource in oil-shale beds or methane hydrates mentioned in Chapter 9 is as large as some claim, the price of gas will come down and the situation change. It is too soon to say how big the usable gas reserves really are.

Very few people under the age of 60 have ever seen a lump of coal. Until roughly 1950 many homes were heated with coal and those of us who are old enough remember what it looks like, and remember with little pleasure how to run a furnace. I can still remember the sound of coal hissing and rattling as it slid down a metal chute into the family basement, and the relief my father and I felt when he converted our heating system to an oil burner and we no longer had to shovel the stuff into the furnace and take out the ashes.

The old 33% efficient coal plants mainly use powdered coal as a fuel, a different form of coal from what slid down into my basement.

---

[1] A reminder; gas gets half of its energy output from turning its hydrogen into water with no greenhouse gas emissions whereas coal gets all its energy from turning carbon into CO_2. Add to this the better efficiency of a gas plant compared with a coal plant and conversion of an old coal plant to a modern gas plant reduces emissions threefold for the same electricity output.

The lumps of coal are ground into something as fine as talcum powder and blown into the furnace with the correct amount of air to assure proper burning, and generate the steam used to run the electric generators. A new coal power-plant technology called Ultra-Supercritical (USC) uses the same powdered-coal fuel but reaches a higher efficiency (about 43%) by running the steam system at higher temperature and pressure than the standard plants. The technology is new, and there are not many of these plants around. USC technology has a slightly older brother called Supercritical (SC) that has an efficiency of 38%, and there are more of these. The newest technology is called Integrated Gasification Combined Cycle (IGCC). Here the coal is first turned into a gas which then runs a combined cycle generator like those used in the best of the gas-fired power plants. While the potential efficiency of the combined cycle part of the system is very good, the gasification process uses lots of energy and the overall efficiency is more like that of the SC plant. The reason for interest in IGCC is that it is easier to capture the carbon dioxide if it is to be captured and stored away (see section 10.2)

As the price of coal rises there will be a switch to these newer technologies with a resulting reduction in emissions for the same electrical output. However, left to present regulations, the incentive is not large and any transition will be very slow. If new coal plants are to be built in large numbers, at the very least regulations should require that they be high efficiency. What is needed to move to more efficient generation is what the economists recommend, a price on emissions as discussed in Chapter 7, which makes disposal of waste part of the cost of doing business. As long as the world's atmosphere is regarded as a free dump for greenhouse gases, utilities will continue to build the cheapest plants. If they had to pay for the emissions, the situation would change in a flash.

10.2 PRICING CARBON EMISSIONS: CARBON CAPTURE AND STORAGE

If there were a way to arrive at an appropriate price for emissions, a price could be set and the incentives to produce power plants with low or no emissions would soar, making the move from a coal-dominated to a lower-emission electricity sector occur much faster. There are many ways to set such a price. One is to set it such that your favorite carbon-free power source becomes less costly than the high-emission ones we use now. Another way is to see how much it would cost to eliminate the emission from our present coal-based system and include that as a fee paid to the government if you emit the greenhouse gas – no emissions means no fee to be paid. The technology being investigated to eliminate the emissions is called carbon capture and storage (CCS).

The basic idea of CCS is to capture the greenhouse gas emissions from our conventional power plants and put them away somehow so that they do not add to the atmospheric load of greenhouse gases. There is an IPCC Special Report on CCS published in 2005 that gives the details [21]. Its Summary for Policymakers is quite readable. Another good, but more technical, reference is a 2007 report from the Massachusetts Institute of Technology entitled "The Future of Coal."[2] Cost estimates from the IPCC report have large uncertainties since they have to consider conditions all over the world while the MIT report focuses on the United States. I will use the MIT numbers for costs.[3]

CCS is used today in limited applications. One example is a Norwegian natural gas field in the North Sea that has a large amount of CO_2 mixed in with the gas. The CO_2 is separated at the well site and re-injected underground into a local reservoir. They inject about 3000 tons of CO_2 per day and estimate that they can put away up to 20 million tons. Another example is enhanced oil production from wells where injecting CO_2 increases the pressure in the oil reservoir, pushing out more oil to be recovered. In this case the CO_2 comes from other industrial processes and is typically sent by pipeline to the oil field where it is pumped underground. Both of these examples are much simpler than CCS at a power plant where a very hot gas stream has to be treated, but they do give an experience base, though a small one, on which to base part of the cost analysis. There is a scale-up issue. A one-gigawatt coal-fired electrical plant produces in a few hours what the Norwegian gas field produces in a day, and there are many thousands of such coal-fired generating plants.

The problem for a power plant in the carbon capture part of CCS is the separation of the CO_2 from the much larger amount of nitrogen in the gas stream. Air used to burn coal is 80% nitrogen and only the 20% that is oxygen combines with the fuel to make CO_2. Two processes are being investigated: separation before or after combustion. In the

[2] http://web.mit.edu/coal
[3] Both the IPCC and MIT reports focus on disposing of the CO_2 as gas injected underground or under the sea. Another process in the research phase looks at the possibility of combining it with other elements into carbonate rock.

first case, the oxygen is separated from the nitrogen in air, and only the oxygen goes into the combustion chamber and only pure CO_2 comes out. It can be cooled, compressed, and pumped off in a pipeline to a storage site. In the second case the hot mixture of CO_2 and nitrogen is cooled and passed through a chemical process that absorbs the CO_2, and the nitrogen is sent back into the atmosphere. The CO_2 absorber is then reheated, the CO_2 comes out, is compressed and sent on its way. In either case the carbon capture process uses lots of energy, and the MIT study estimates that the CC part of CCS lowers the overall generating efficiency by nine percentage points; that is a reduction to 34% efficient from the 43% typical of a USC plant.

The MIT study estimates that the cost of electricity would increase by about 3 cents per kilowatt hour (kWh) because of the carbon capture process. Although they did not estimate the cost of carrying it away and pumping underground, I would guess that would add another 1 to 2 cents per kWh, for a total CCS cost of 4 to 5 cents per kWh. Since one kilowatt hour of electricity from coal produces about one kilogram of CO_2, at that rate CCS costs $40 to $50 per tonne of CO_2, or around $160 per ton of carbon. The European Union has a carbon emissions market that has functioned for several years. Prices have been quite volatile and as of May 2009 are around $20 (€15) per tonne, well below my estimate of capture and sequestration costs.

Two ways to internalize the cost of emissions are being discussed that if done would dethrone coal as the low-cost power source and turn industry to other fuels that emit less or no CO_2. One proposal is to impose a simple fee on emissions. The other is called Cap and Trade and would limit the total emissions from all sectors. These are discussed later in the policy chapters.

10.3 DOES WHAT GOES INTO STORAGE STAY THERE?

My skepticism about CCS is less about the capture technology than about the ability of the storage systems to keep the CO_2 out of circulation for a long time. Two scenarios are being discussed. One puts the CO_2 in the deep ocean, transporting it there either by pipeline or as a liquid in a ship. This one doesn't work. The other puts it under the surface in depleted oil and gas reservoirs or in what are called deep saline aquifers. These contain mineral-laden waters deep underground that are not connected to fresh water supplies. This one might work.

The issue is, does the CO_2 stay where you put it? For the oceans we know the answer and it is no. Surface waters in the oceans are slowly carried to the deeps and deep waters are slowly transported to the surface. How fast the process runs depends on the depth. The IPCC Special Report estimates that if we deposit 100 years' worth of CO_2 produced in this century at a depth of about 2500 feet in the ocean, about 50% would have come back out by the end of the next century. Deeper is better, and if it were put at a depth of 5000 feet only 25% would have come back out.

Storage underground on land raises two issues. The first is the same as for the oceans – does the CO_2 stay where you put it? The second is the capacity of the underground sites. The IPCC Special Report estimates that, worldwide, depleted oil and gas reservoirs can store between 675 and 900 billion tonnes of CO_2 while the deep saline aquifers can store between 1000 and 10000 billion tonnes. The business-as-usual scenario would give about 4000 billion tonnes for the total emissions in this century, and that much is a tight fit for what we know of the potential storage sites. In addition, the sites are not uniformly distributed around the world. There seem to be lots of deep saline aquifers under the United States and few under China, the two largest emitters.

We know that the oil and gas reservoirs did not leak before they were exploited. If they had, the gas would be gone and the pressure in the oil reservoirs would also be much lower than it is found to be. However, the reservoirs have had many holes punched in them, and all those holes would have to be plugged; probably not a big problem, but a concern. We do not know about the long-term behavior of the saline aquifers. The leaks of concern in those are not like the catastrophe that occurred at Lake Nyos in the Cameroon in 1986 [22]. (Nyos, a lake with huge amounts of CO_2 dissolved in its cold bottom waters from volcanic activity, released the gas in a rush, creating an asphyxiating cloud that suffocated most living things within 10 kilometers.) Leaks from CO_2 reservoirs will be slow. However, given the litigious nature of many of the world's countries (particularly the United States), liability for leaks will surely delay any large-scale implementation of a CCS program.

The CCS option needs to be tested. If it works, the world will be able to continue using coal for some time while newer carbon-free technologies mature, especially important for the developing nations. We know enough from small-scale applications to know that it works in principle, but we do not know if it works at the scale required for a power plant. The US DOE started such a project, but recently canceled it as costs went up and up and up. I always thought this project, called

FutureGen, was misguided. It tried to do too much: produce electricity, produce hydrogen, and demonstrate CCS at an industrial scale. I would separate the parts. An industrial-scale demonstration of CC needs to be done for separation both before and after combustion to learn something about costs and efficiency of the capture process. How much energy does it take and what fraction of the CO_2 is captured in the real world? CCS does not have to be perfect to make a huge contribution to cutting greenhouse gas emissions. At 90% capture efficiency, coal would become a small contributor to climate change and the world would have much more time to develop the technologies that will be needed in the long term to get emissions fully under control.

Tests of the deep saline aquifers with carbon dioxide loading don't need a new power plant to supply the gas. We can start with CO_2 from any source (a lot is needed) and begin to learn what happens as the system becomes more acidic with CO_2 loading. No one seems yet to be doing that.

10.4 SUMMARY AND CONCLUSION

Coal is the mainstay of our electricity production and is not going away soon. Indeed, coal use is growing. In the United States over 150 plants are in the planning or construction phase. China and India are adding plants even more rapidly (200 in 2008 alone for China), and even Europe as of the time of this writing (early 2009) is considering 50 new ones.

We will have those emissions with us for a long time unless there is some powerful incentive to phase them out. The best incentive is to make other options more attractive than continued reliance on coal. That can be done in several ways, one of which requires that emissions be made to bear some sort of cost. Based on the analyses in the IPCC special report and the MIT report I concluded that CCS costs about $40 to $50 per ton of CO_2. We do not have to know if CCS works to add a fee to CO_2 emissions of $45 per ton and let the emitters work out how to reduce their costs by increasing the efficiency of their power plants, developing CCS systems, or turning to other sources of electricity that would no longer need subsidies to be more economically attractive than coal.

In the spirit of no silver bullets, CCS is worth an industrial-scale experiment aimed at storage in the deep saline aquifers. They have the capacity to store a large amount of CO_2 and therefore CCS has the potential to solve part of our problem. The test should be a combined private–government funded program. FutureGen, begun several years ago but canceled in 2008 when costs greatly increased, was the wrong program because it tried to do too much too soon. If I was running the program I would have had an existing coal plant equipped with the best of the post-combustion CO_2 separations technology to test it out. I also would have supported a test of pre-combustion separation of oxygen from the nitrogen in the air. Finally I would have a program of CO_2 injection into the deep saline aquifers to see what happens to it. The latest congressional budget has funds for CCS, and it will be interesting to see what kind of test is supported.

Conversion of old coal plants to modern gas plants reduces emission to one-third of the original coal plant emissions for the same electric power generation. Part of the reduction comes from the fuel switch and part comes from the higher efficiency of the gas-fired power plants. This should be encouraged, but somehow it doesn't seem to have the emotional attraction of renewable energy, though it is much less costly, and can be done on a very much larger scale than solar power.

More emission-free electricity can come from nuclear power plants or from the renewables. These are the subject of later chapters.

11

Efficiency: the first priority

11.1 INTRODUCTION

There are many recent studies by governments, non-governmental organizations, and the private sector that all come to the same conclusions:

- Improving energy efficiency is the cheapest and easiest way to reduce greenhouse gas emissions;
- Energy not used reduces imports, emits no greenhouse gases and is free;
- The transportation and building sectors use far more energy than is necessary;
- The total cost to the economy as a whole of most of the improvements is negative: we save money.

In this chapter I look at what might be done to improve energy efficiency in two of the three sectors of the US economy: transportation and buildings. The third sector, industry, has to have each process looked at separately and that is too big a job for this book.

Improving energy efficiency in buildings reduces electricity demand, thereby reducing fossil fuel use in generation, and reduces fossil fuel use for heating as well.

Increasing the efficiency with which energy is used in the transportation sector does more than reduce greenhouse gas emissions; it also reduces the imports of large amounts of oil to fuel that sector and thereby also reduces the export of the large amount of money that goes with those imports. With the recent increase in the price of oil, even those few people who do not believe that cutting greenhouse gas emissions is important to reduce the danger of climate change agree that reducing oil imports is important and have become allies in a move toward a more efficient economy.

Given those conclusions, it may be surprising that so little has been made of the opportunities. In a market economy like those of most of the developed countries of the world, consumer demand forces manufacturers to comply with that demand in order to be competitive, or manufacturers see an advantage and work to convince the customers to buy what the manufacturers want to sell, or the government sees some national importance that makes it force efficiency on the society. Until recently the price of energy was so low that there was little if any consumer pressure for energy efficiency, manufacturers had little incentive to invest in better efficiency technology, and there was no concern on the national level because of any economic drain on the economy from energy prices.

We have been through a major problem before with the cost of energy. In the early 1970s the Arab members of OPEC imposed an embargo on oil shipments to Europe and North America, and the price of gasoline shot up. In 1979 the Iranian revolution deposed the Shah; the United States gave the Shah sanctuary; and the Iranian students seized the American Embassy, holding the staff hostage for nearly two years. Oil prices remained high for over a decade. Long lines at gasoline stations and high prices stimulated the government to action, and the action was effective. I used a version of Figure 11.1 in Chapter 6 as an illustration of the overall long-term decline in energy intensity, a measure of the efficiency with which energy is used in society. This time I use a version of the figure that includes what happened to fuel prices. The high price of oil stimulated a move to efficiency, and energy intensity dropped much faster in the 1970s and 1980s than the historical average. What was a decline of 1% per year became 2.7% per year during the period of what was called the oil shock.

The first Corporate Average Fuel Economy (CAFE) standards were introduced in the United States in 1975 as a result of the 1973 oil embargo and when fully effective in 1985 resulted in a doubling of the average miles per gallon (mpg) of the auto fleet from 14 mpg before the oil shock to about 27.5 mpg for cars and 21 mpg for light trucks. The first appliance energy standards were introduced in California and resulted in much more efficient appliances; for example, refrigerators are larger today than in those days but use much less energy (federal standards were not introduced until 1989 but manufacturers used the California requirement nationwide). But the price of oil dropped precipitously in 1986 and with it came a drop in efforts to improve efficiency. Supply was going up as more countries began to export oil, and demand was dropping because of increasing

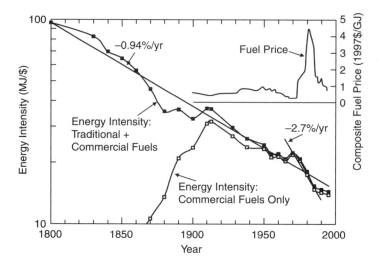

Fig. 11.1   Energy intensity versus time. This version of the figure
for energy intensity versus time includes what happened to fuel
prices in the 1970s and 1980s. It shows the coupling between a price
shock and a move toward higher efficiency and, hence, lower energy
intensity. (Courtesy of Prof. S. Fetter, University of Maryland. © Steve
Fetter, *Climate Change and the Transformation of World Energy Supply*,
Stanford: Center for International Security and Cooperation, 1999)

efficiency. OPEC decided that they were playing a losing hand and
lowered their prices.

This time it is different. We have a real supply crunch as well as
global warming to deal with. The developing world, particularly China
and India, has begun to move up the economic ladder. Demand for
fuel has increased greatly: as of 2007, according to the EIA, demand
for oil is going up each year by about one million barrels per day and
has reached 85 million barrels per day, the price of oil has gone up,
and demand for coal and natural gas has resulted in large increases
in their prices, too. I am writing this in December of 2008 when the
price of oil has come down to below $50 per barrel from its summer
peak of nearly $150 per barrel. When I was in Iran in the spring of
2008, their officials felt the proper price based on demand was about
$80 to $100 per barrel, and the high price was due to speculation.
Thanks to the greed and incompetence of the financial sector of the
world economy and its regulators, we are now in what looks like a
worldwide recession and the bet is that demand for oil will decrease
as the world economy slows. But the growth of the developing countries will continue, and the demand part of the economic equation

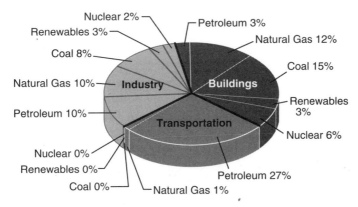

Fig. 11.2   TPES by fuel and sector. (Courtesy of the American Physical Society, Ref. [24])

will pull energy prices up again sometime soon. I can only hope that this temporary relief will not result in abandoning the move toward efficiency. Remember please, this time is different from the last time. Growing demand is forcing price increases and we have to deal with global warming too.

Figure 11.2 shows how energy is used in the US economy.[1] The transportation sector uses 28% of all the primary energy and 95% of it is oil-based. The buildings sector uses 39% of our total primary energy and a large fraction of that is used to generate the electricity consumed in commercial and residential buildings. The supply is diverse because electricity generation uses so many different primary energy sources (renewables in the figure are dominated by hydroelectric power and biomass). The US Environmental Protection Agency (EPA) allocated 33% and 38% of greenhouse gas emissions from energy use to the transportation and buildings sectors respectively. Reducing the energy and emissions numbers is what the rest of this chapter is about, and large reductions are possible at little if any cost to the economy.

In what follows I discuss end-use energy efficiency and include all the energy that goes into some application (the primary energy), as well as looking at how much energy is used compared to the minimum to do the same job.

---

[1] I recently chaired a major study of efficiency in the US transportation and buildings sector. Some of the material in this chapter comes from the report which is available at http://www.aps.org/energyefficiencyreport/index.cfm . It goes into considerable detail for those interested.

Primary and end-use energy efficiency

Primary energy is what goes into a process or a product and includes what is used to make secondary forms of energy. For example, electricity is not a primary form of energy. It has to be generated in some fashion using some kind of fuel. Electricity at the wall plug in the United States on the average contains only 31% of the primary energy used to generate it. Totaling up all of the energy used in a building has to include the energy used to generate the electricity used, if we are to get at the primary energy use.

Suppose I have two houses heated by gas heaters of 90% efficiency, but one house is well insulated while the other needs twice as much gas to keep it warm. The heaters have the same efficiency, but the well-insulated house uses half the energy and so in end-use terms it is twice as efficient.

Plug-in hybrid cars will soon be on the road, getting part of their fuel by charging their batteries from the electric power grid, and part from running on gasoline. The miles per gallon of gas will go way up, but the electrical energy needs to be counted too. The primary energy used to generate electricity plus the energy in the gasoline need to be counted in getting at the total. If we ever get hydrogen-powered vehicles, the energy required to make the hydrogen has to be counted to get the primary energy requirement and the overall efficiency.

11.2 TRANSPORTATION

I have some direct experience with efficiency in the transportation sector thanks to my wife. She was one of the few who had a General Motors EV-1, their all-electric car, which she received on one of those important decadal birthdays. It could (and she did) out-accelerate a Porsche up to about 30 miles per hour thanks to the very high torque of electric drive at low speeds. With night-time charging at off-peak rates, it cost her only about one cent per mile to drive compared with the 10 cents per mile my car cost to drive when fuel was only $2.00 per gallon. Since electric drive is much more efficient than the standard gasoline engine (more on this later), she was two to three times as efficient in primary energy terms, too. Sadly, General Motors took back all of the EV-1s and crushed them, something Rick Wagoner, GM's former Chairman and CEO, now says was one of his worst mistakes. My wife threatened to elope to Mexico with her car, but I did manage to talk her out of it. She now drives a Prius, and looks forward to a future all-electric.

The transportation system runs on oil. On an average day, the United States uses 20 million barrels of oil (one barrel contains 42 gallons), two-thirds of which is imported. Transportation uses 70% of the oil, and when oil was $145 per barrel the US public spent $1.4 million per minute on oil for transport and paid $900 000 per minute for imports, a considerable amount of which went to countries like Venezuela whose governments did not like the United States very much. Now that the price of oil is down to a mere $50 per barrel, the United States spends only about $480 000 per minute and pays only $310 000 per minute for imports. Incidentally, the United States is the world's third largest producer of oil after Saudi Arabia and Russia, although domestic production is going down.

After refining, the products coming from a barrel of oil are [23]:

- 47% gasoline
- 23% diesel fuel and heating oil
- 10% jet fuel
- 4% propane
- 3% asphalt
- 18% other petrochemicals for things ranging from chewing gum to plastics.

The total is more than 100% because the refining process adds material to the original oil so the total volume of output is about 105% of the input oil. The breakdown energy consumption by the elements of the transportation system in the United States is as shown in Table 11.1.

This section looks at light vehicles: cars, minivans, SUVs, and pickup trucks; which use the largest part of transportation fuels, equivalent to 9 million barrels of oil per day. The other sectors are important, too, but the biggest is the light-vehicle sector and that is the focus here.

Table 11.1 *Transportation energy consumption by mode (2005)*

Light vehicles	63%
Heavy-duty road vehicles	17%
Aircraft	9%
Water transport	5%
Pipeline operation	3%
Rail transport	2%
Motorcycles	1%

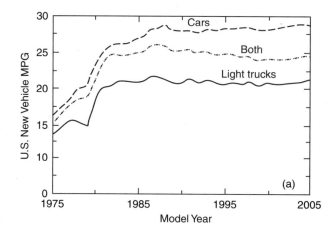

Fig. 11.3a   US miles per gallon. Fuel economy of US cars and light trucks, 1975–2005. (*Sources*: US Environmental Protection Agency, National Highway Traffic Safety Administration)

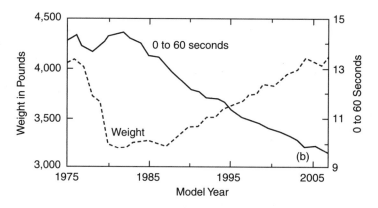

Fig. 11.3b   Vehicle weight and acceleration. Vehicle weight initially decreased to help meet the new standards, but has increased ever since. (*Source*: US Environmental Protection Agency, 2007)

Fuel economy yesterday, today, and tomorrow

Before looking at what can be done to improve efficiency it is interesting to look at what has been happening since the first US corporate average fuel economy (CAFE) standards went into effect. Figure 11.3a shows the average fuel economy of the US light-vehicle fleet and Figure 11.3b shows the average weight and performance of that fleet.

Figure 11.3a shows that mpg changed very little after the CAFE standards came into full force in 1985. Figure 11.3b, however, shows a dramatic change in the weight and acceleration of vehicles. The weight

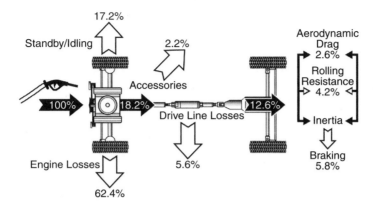

Fig. 11.4   Where does the energy go? How energy flows for a vehicle powered by an internal-combustion engine. The diagram shows the energy uses and losses from a typical vehicle. (*Source*: fueleconomy.gov)

first went down to help meet the standards (a 10% weight reduction results in a 7% mpg increase) and then went up steadily from 1987 to 2007. Times for acceleration from a stop to 60 mph went dramatically down over the same period. To do this without reducing fuel economy required a major increase in engine efficiency. The auto makers knew perfectly well how to do it, but when the price of oil crashed down in 1986, customers no longer demanded more fuel economy, so improved efficiency went into performance and increasing the size of vehicles rather than improving fuel economy. Only government regulations held mpg steady. If you look hard you can find many articles in the late 1980s arguing that it was un-American to restrict the kind of cars people could buy by restricting fuel economy. Fortunately, the government did not listen, for if they had US oil imports would be much larger than they are today.

The average energy efficiency of a typical car with a gasoline-powered internal combustion engine (ICE) is very low. Figure 11.4 is from the US Environmental Protection Agency and shows where the energy in the gas tank goes.

What is called the "tank-to-wheels" efficiency is the fraction of the energy in the gasoline that actually moves the car. In the standard ICE car it is a low 12.6%. More than 60% of the primary energy is lost in the engine to friction and heat, 17% is lost to idling, and 2% goes to operate accessories such as lights, air conditioning, and radios. Only about 18% of the primary gasoline energy is delivered to the output of the engine and from there a further third is lost in getting through the transmission, drive train, and differential (there is almost no difference in the loss between front-wheel and rear-wheel drive cars).

The 12.6% that actually moves the vehicle goes to overcoming aerodynamic drag and rolling resistance, and to braking losses when slowing.

Aerodynamic drag depends on the design of the vehicle and the speed; the 2.6% loss to aerodynamic drag is an average. Rolling resistance depends on tire design and pressure. Under-inflation can cost considerably in fuel economy (sorry to say there are ignorant politicians who ridicule this notion). The remaining part of the energy delivered to the wheels goes to accelerate the vehicle to its operating speed and is lost to heat in the brakes when speed is reduced. Typically, fuel economy is lower in city driving than in highway driving. The main contributor to the reduction for city driving is the more frequent stops and starts. Each time a vehicle is accelerated to operating speed a certain amount of energy is used. Each time it is stopped that energy is thrown away (except in a hybrid like the Prius). While it takes more energy to accelerate to highway speed, the more frequent stops and starts in city driving are the main factor resulting in a lower fuel economy.

The Toyota Prius is an example of a hybrid that uses two drive systems: an ordinary ICE, and an auxiliary electric motor and battery system that recovers energy lost in braking and uses it again in a system much more efficient than an ICE working alone. The secret of the hybrid is that electric drive is about 90% efficient in delivering energy from a battery to the wheels, compared with the ICE 12% efficiency. You do not have to recover all of the braking loss to make a big difference. In addition, the hybrids turn the engine off when idling for a long time and reduce losses there too. My wife's Prius gets about 45 mpg in her city driving, roughly double what an ICE would get. It can use either or both engines at any one time.

What I think of as the ultimate hybrid is sometimes called a serial hybrid. Here, the ICE only drives an electrical generator that either charges the battery or helps drive the electric motor. This allows the ICE to operate most of the time at its optimum efficiency, getting even more of an improvement than systems like that used in the Prius. General Motors' Chevy Volt, a plug-in hybrid that will be discussed later, operates as a serial hybrid when running on gasoline.

The new US CAFE standard for all vehicles is 35 mpg by 2020, but there is no reason to stop there (see Technical Note 11.1 on how to define average fuel economy over a fleet of different vehicles). The American Physical Society's 'Energy Future: Think Efficiency' study [24] recommends that the government require 50 mpg by 2030 for vehicles that operate only on liquid motor fuels, though they do not say how to do it. It could be through a new CAFE standard or something like a fee for vehicles that do not meet the standard and a rebate for those that better it (sometimes called a fee-bate). Shoppers love bargains, and a fee-bate might do well at stimulating demand for more efficient vehicles. Technologies to move beyond 35 mpg are already in the works or exist. Here are a few examples (see Ref. [24] for more detail).

- Diesel engines in Europe have about 30% better fuel economy than gasoline engines (operation is at higher compression ratios, and diesel fuel has more energy per gallon than gasoline). According to *The Economist* magazine (October 24 2008), Volkswagen has a fleet of hybrid diesels running in Germany that get over 90 mpg.
- HCCI (Homogeneous Charge Compression Ignition) engines run on gasoline but have the efficiency and fuel economy of diesel engines; they are being developed by most car manufacturers.
- Every 10% weight reduction produces a 7% improvement in fuel economy.

Weight and safety

There is a concern that reducing vehicle weight will result in more injuries in crashes. However, crashworthiness is largely in design, not in size or weight. A report by the International Council on Clean Transportation [25] (ICCT) breaks safety into three components: crashworthiness, crash avoidance, and built-in aggressivity (a wonderful term that I had not heard before).

- All passenger vehicles are crash tested and rated for crashworthiness using instrumented dummies to simulate what would have happened in real life. You do not have to be big to be good. The very small SmartCar made by Mercedes has a high crashworthiness score.
- Crash avoidance is related to both technology and design. Antilock brakes allow a panic stop without an uncontrolled skid, while a high center-of-mass, like some SUVs have, increases the tendency to roll over. The first improves avoidance ability while the second increases risk.
- Aggressivity is more than just looks. If you look at the SUVs and pickup trucks on the road used as passenger vehicles, you will see that some have low bumpers that match those on cars while some have high ones. The high one will override the bumpers of

a car in a collision and thereby do much more damage. There is no standard that prevents the sale of these killer vehicles. They have a high aggressivity.

The ICCT report concludes that there is no direct relation between safety and size as long as you are talking about vehicles that do not differ in weight by too much. There is no doubt about who will come off worst in a collision between a passenger vehicle and a large highway truck.

Plug-in hybrid and all-electric vehicles

The success of hybrid vehicles like the Toyota Prius has led to an effort to go much further with the technology by starting with a battery that has enough charge to get the driver through a short trip without using any gasoline. These are called plug-in hybrid electric vehicles or PHEVs. In the San Francisco Bay area where I live there are several auto shops that will convert a standard Prius to the plug-in variety by installing extra batteries to get some reasonable range, typically 10 to 15 miles. For longer trips the vehicle reverts back to the normal hybrid mode running on gasoline with the efficiency gains typical of a hybrid. If you never drive more than 10 to 15 miles before recharging the batteries you will never use any gasoline. If you typically drive 20 to 30 miles before a recharge you will do half on the batteries and half in the normal Prius mode, but the 45 mpg of the standard mode will get you 90 miles per gallon as a PHEV (that does not account for the primary energy used to generate the electricity). There is no standard mpg for a PHEV – what you get depends on your driving pattern. Of course you do have to pay for the electricity.

The major auto manufactures are planning the introduction of their own plug-in varieties. Toyota has announced that it will bring out a PHEV version of the Prius with an electric range of about 10 to 15 miles in 2010 (plug-ins are designated by their range on batteries alone so the Toyota would be a PHEV10 or PHEV15). General Motors will introduce its PHEV40 Chevrolet Volt in 2010. Other manufactures will be entering the field as well, and all-electric vehicles are about to make a comeback, first for short trips and then for longer trips as battery technology improves.

If the batteries are good enough, the potential impact on oil use in the light vehicle transportation sector can be huge. Figures 11.5 and 11.6 show the potential. Figure 11.5 is based on data from the

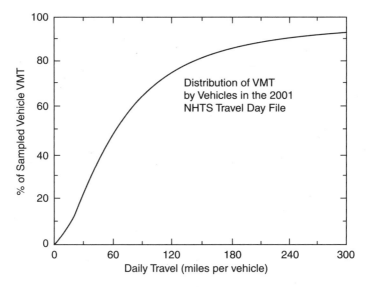

Fig. 11.5   On the road. Percentage of sampled vehicle miles traveled (VMT) as a function of daily travel. (*Sources*: [26, 27, 28]. Credit: Argonne National Laboratory, managed and operated by University of Chicago Argonne, LLC, for the US Department of Energy under Contract no. DE-AC02–06CH11357)

National Highway Traffic Safety Administration (NHTSA) and shows the percentage of light vehicles that travel less than some number of miles per day [26, 27, 28]. For example, 25% of all vehicles travel less than 30 miles per day, and about 55% travel less than 60 miles per day. Figure 11.6 looks at the numbers from a different perspective: what percentage of all vehicle miles traveled would be traveled on electricity in a fleet of PHEVs with a certain range on electricity alone [26, 27]. Even if you travel more than the range allowed by the batteries for all electric operation, the first part of your trip is on electricity. Looked at this way, if all of the light-vehicle fleet was PHEV40s, 60% of vehicle miles traveled (VMT) would use no gasoline, while if they were PHEV100s 85% of travel would use no gasoline.

Toyota seems to be starting its PHEV program with the same kind of nickel-metal-hydride batteries used in the conventional Prius while they work to develop advanced lithium-ion batteries that pack more energy and power into the same space. General Motors has been working for several years on advanced batteries. More than just an improvement in capacity is required. The Prius of today runs on a previous generation of batteries, which only allow about 20% of the energy in the battery to be used. The restriction on how much energy can be drained from the battery before a recharge is to improve battery life. It is important to develop batteries that can stand deep discharge or the weight and size of the battery pack goes up excessively. It is not what you can store, but what you can use that counts. The advanced lithium-ion battery for the Volt will be expensive at first, and will take some time to develop into a compact, low-cost form, but once it gets on the road it will go through the normal process of continuous improvement.

The impact of a full fleet of PHEV40s on demand for gasoline is huge. If all the light-vehicle fleet were PHEV40s, gasoline consumption would decrease dramatically. In the United States the 9 million barrels (bbl) of oil per day used in the transportation sector would drop to 3.6 million bbl per day and oil imports would drop steeply. However, don't hold your breath until the new day dawns. The PHEVs are new, and battery development will have to proceed for a few years until they have the capability to handle the large-car end of the motor fleet. Also, cars turn over slowly. A census of all the vehicles on the road today will find some as old as 15 years, and it will take 15 more years until the newest one on the road today is retired.

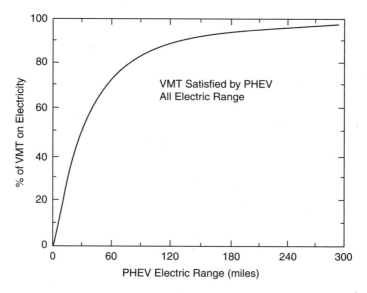

Fig. 11.6   Electric-powered driving. Fraction of VMT driven on electricity as a function of the PHEV electric range. (*Source*: [26, 27]. Credit: Argonne National Laboratory, managed and operated by University of Chicago Argonne, LLC, for the US Department of Energy under Contract no. DE-AC02–06CH11357)

All-electric vehicles need much more battery capacity than the plug-ins if they are to meet the needs for a large part of the population. There have been all-electric vehicles before, such as the General Motors EV-1 that my wife had, and the Toyota RAV-4 that were very popular with their users. The evolution of battery technology is worth noting. The first EV-1 was introduced in 1996 and had a range of only about 60 miles. The second version was introduced in 1999 and had an improved nickel-metal-hydride battery and a range of about 120 miles. The new Tesla sports car uses lithium-ion batteries and has a range of 240 miles (it costs over $100 000). Battery technology will continue to improve.

The US goal for general-purpose all-electric drive is a range of 300 miles, and achieving that will require much better batteries than exist today. A 300-mile range will require a battery capacity of seven and a half times that of a PHEV40, a formidable development task. However, several all-electric cars will be on the market in the next few years. The TESLA sports car already mentioned has a range of about 240 miles. A joint venture by Nissan Motors and Renault plans to introduce a small electric car with a range of about 100 miles and a top speed of 75 mph. It will be a while before all-electric drive vehicles are available for the mass market.

One of the problems with all-electric drive is the time it takes to fully charge a depleted battery. Even with a 220-volt electrical source, charging time will be many hours with present battery technology.[2] This is not the sort of time a driver is likely to want to spend having lunch while his battery is recharged during a long trip. A new company has announced a venture to have a battery exchange program – drive up with your discharged battery and have a full one installed in a short time. There is an advantage that PHEVs have over all-electric: very long trips can be completed on gasoline. If for example, all light vehicles were PHEV100s, batteries would only need to be two and a half times as large as PHEV40s, 70% of all daily driving could be accomplished on electricity, and gasoline use would decrease by 85% – not bad if you are for energy independence.

The electric power grid

If the economy moves to PHEVs or all-electric vehicles, the electricity has to be produced somewhere and delivered from there to the

[2] The TESLA car uses 340 watt-hours per mile according to the company. In 200 miles of driving it uses a total of 68 kilowatt-hours. Recharging in 10 minutes would require a 200 kilowatt power source, in 1 hour a 68 kW source, in 4 hours

vehicle. There is a potential problem with the electrical power grid, however, because the amount of power needed for a large number of PHEVs or all-electric vehicles is large. A preliminary industry estimate for electric vehicles is that on the average they will use about 340 watt-hours per mile traveled. In the United States, vehicle miles traveled (VMT) in the year 2008 was about 3 trillion miles. That implies that if all vehicles were PHEV40s it would take about 200 gigawatts (GW) of electric power to charge them (8-hour charging time) and about 350 GW for all-electric. That much power is not now available during the day, but is available at night when electricity demand drops.

Peak electricity demand everywhere is in the daytime. In the United States there is 200 GW or more to spare at night. But not all the extra capacity is available where the cars are, and so the electric power distribution grid will need modification. (It needs it anyway to be able to ship around electricity generated from wind and from solar power sources; the big potential sources in the United States are in the Great Plains and the southwest, while in the United Kingdom the wind sources are in the north whereas the largest demand is in the south.) The California Energy Commission has said the state can handle a million plug-in hybrids with its present grid (I have no information abut Europe). If market penetration is as slow as it has been for conventional hybrids there is plenty of time for an upgrade to the electrical distribution system. The total number of hybrids sold in the United States from their introduction in 1999 to the end of 2008 is only about 400 000. High gas prices will make the introduction of plug-ins faster. The price of oil has gone down as of this writing but it will go up again when the world economy recovers. I can only hope that the drive to improve efficiency will not stall as it did in 1986 when oil prices fell after the first oil shocks.

Other fuels

Other fuels have been discussed as possible substitutes for gasoline or electricity from the power grid. The main candidates are ethanol, natural gas, and hydrogen. Ethanol made from corn is what the US program mandates now, though more advanced ways of making ethanol are in the research phase. Ethanol is discussed in detail in Chapter 14 on biofuels. Here I only say that corn ethanol takes about as much

a 15 kW source. A typical house has an electrical feed of 20 kW. A 10 minute recharge is a fantasy.

energy to make as is in the gasoline it displaces, results in the emission of nearly as much greenhouse gas as gasoline, and has driven up the price of food. Corn ethanol is a subsidy for agribusiness, not a route to greenhouse gas reduction. There are promising new biofuels under development, but none have yet reach the stage of commercial-scale production.

Natural gas can be used directly in an ICE. Honda makes the Civic GX which costs about $25 000 and is rated by the California Air Resources Board as an "advanced technology partial zero emission vehicle". In residences, natural gas is widely used for heating and cooking. The natural gas line can be fitted with a take-off and compressor needed to compress gas and fill a car's tank. People who have the GX simply hook them up to the gas line at night and drive them out in the morning. There are also some refueling stations where you can get the tank recharged. The GX is popular in places where natural gas is cheap. Since natural gas produces lower greenhouse gas emissions than does the amount of gasoline with the same energy content, emissions are reduced. However, if these vehicles achieve wide use, natural gas imports into the United States, Europe, China, India, etc. will increase unless much more gas is found in those countries. It is possible to make engines that can run on either natural gas or liquid fuels. To my knowledge, none are on the road.

Hydrogen is another potential source of fuel that has received much publicity. There is a huge amount of it, but hydrogen is so reactive that all of it is tied up in various chemical compounds including water and natural gas. There are no hydrogen wells. The first problem for hydrogen is the energy it takes to separate the hydrogen from the other elements to which it is bound. The second problem is how to use it. It is possible to burn hydrogen in an ICE but that is no more efficient than burning gasoline; all the losses shown in Figure 11.4 will happen and in addition more energy is lost to producing the hydrogen. The thrust of the development program is to use what is called a fuel cell to make electricity from hydrogen to run an electric drive system. The two alternatives receiving the most attention are generation of hydrogen on board the vehicle from natural gas (or perhaps methyl alcohol), or generating the hydrogen outside the vehicle and storing the hydrogen itself on board. Most of the effort to date has been on the systems with external hydrogen generation.

The vision of the advocates of hydrogen fuel is that wind- or solar-generated electricity is used to produce the hydrogen; the hydrogen is moved to the fueling station through pipelines and then used on board the vehicle to generate electricity in a fuel cell. You might well ask why we should go through all this when we have a perfectly good way to move the electricity itself. I have asked this question myself and have never gotten a good answer. Setting that aside, there are some basic science questions that need to be answered before hydrogen can be considered as a large-scale fuel source. The first of these has to do with the fuel cell itself. Currently, efficiency of the fuel cell is low. The target is about 65% for turning hydrogen into electricity, but the best that seems to be doable now is about 50% and at full load none so far does even that well.

The second problem has to do with the platinum catalyst used to speed the reaction that leads to electricity production in the fuel cell (platinum is also used in the catalytic converters on all cars to reduce emissions that lead to smog production). Current fuel cells use about 2 ounces of platinum, and if all the cars produced each year in the United States alone were equipped with these fuel cells, the entire world production of platinum would not be not enough to meet the demand. Until a better system of catalysis is developed, hydrogen fuel-cell vehicles cannot be more than a niche market. The fuel cells should be sent back to the basic research laboratories.

There are other problems as well. Hydrogen production efficiency is below target, a new pipeline system is needed to distribute the gas, and a better on-board storage system is needed. I have never understood why hydrogen has gotten so much R&D money and batteries so little.

11.3 BUILDINGS

We each spend more than 90% of our time indoors – working, shopping, eating, entertaining, and sleeping – in buildings that in the United States are collectively responsible for 39% of primary energy consumption, and 36% of greenhouse gas emissions.[3] The building sector, like transportation, uses much more energy than it needs to, but fixing the problem is much more difficult than for transportation. Transportation is dominated by a small number of large producers; buildings are a mirror image – a large number of small producers. In

---

[3] A reminder to the reader – I always discuss efficiency in terms of primary energy consumption. In the building sector the biggest difference is in the energy used to produce the electricity used in a building. On the average only 31% of the primary energy used to generate electricity emerges at the wall plug.

the United States, transportation, safety, and mileage standards are a Federal responsibility; in buildings, construction codes and standards are mainly a state responsibility, making it 50 times more difficult to get anything done. Only in appliances, heating, ventilation, and air conditioning systems does the Federal government have a say and that say has resulted in large increases in efficiency.

The national stock of buildings grows slowly and turns over slowly. Typical lifetime for building structures is about 100 years, while the systems inside turn over every 20 years. The total net number of buildings (new minus demolished) in the United States grows only by 1% to 2% per year. To make a major impact on efficiency requires not only that that new buildings be efficient, but also that cost-effective retrofits be introduced.

Total energy used in the buildings sector has been growing faster than the population, which should be no surprise because we use energy for many more things today than we did 50 years ago. Figure 11.7 from the DOE's Annual Energy Outlook (2008) for the United States shows that energy use has gone up by four times since 1950 (population has only doubled according to the census bureau), and is projected to grow by another 30% by the year 2030. The total

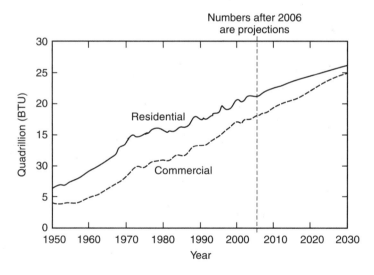

Fig. 11.7   Total primary energy consumptions for buildings. Primary energy use (including that associated with electricity generation) for the residential and commercial sectors in Quad (10^5 BTU). (*Source*: US Department of Energy, Energy Information Administration, Annual Energy Outlook 2008)

Table 11.2 *Primary energy use in buildings*

Residential		Commercial	
Space heating	32%	Lighting	27%
Space cooling	13%	Space heating	15%
Water heating	13%	Space cooling	14%
Lighting	12%	Water heating	7%
Refrigeration	8%	Electronics	7%
Electronics	8%	Ventilation	6%
Cooking	5%	Refrigeration	4%
Wet cleaning	5%	Computers	3%
Computers	1%	Cooking	2%
Other	3%	Other	15%
Total energy (Quads)	**21.8**	**Total energy (Quads)**	**17.9**

energy used today in buildings is 39 Quads (a Quad is a million billion BTU); a huge amount of energy that is just as hard for me to get my head around as it is for the reader. The total primary energy used in the United States is 101 Quads, so just think of it as about 39% of all the energy used in the United States.

A reasonable and achievable goal for the building sector as a whole is that it use no more energy in 2030 than it does today. The projection of a 30% increase in energy use really only says that efficiency remains about the same over the next 20 years. Since the stock of buildings grows by about 1.5% per year and we look 20 years into the future we should expect to use about 30% more energy if we continue as we are. If we can improve overall efficiency by 1.5% per year to match the growth we can achieve zero growth in this sector, and we should be able to do better than that. To improve building efficiency it is useful to start by understanding how energy is used, and it is used somewhat differently in the commercial and residential sectors. Table 11.2 shows the top uses for each sector (from the 2007 Energy Data Book of the DOE).

As of 2005, in the United States there were 113 million residences totaling 180 billion square feet (houses, apartments, and trailers), and 74 billion square feet of commercial space. The top four energy consumers are the same in the residential and commercial sectors, and between them account for 70% and 63% of energy use, respectively. To have no growth between now and 2030 requires that we reduce energy consumption by 12 Quads out of the roughly 50 Quads projected for 2030.

Better building design and more effective use of existing technologies can greatly improve building efficiency without waiting for new inventions. Here are some examples:

- Better insulation can be installed to reduce heat loss in winter and gain in summer (most residential buildings are under-insulated);
- Window coatings are available to reduce heat loss though thermal transmission (heat in and out is in the infrared band and can be blocked while letting all the visible light in);
- Improved furnaces and air conditioners can be built (furnaces are fairly good now, but there is considerable room for improvement in air conditioners);
- White roofs can reduce heat absorption in summer and radiation in winter (they can even help cool the planet by reflecting some incoming radiation back into space just as the ice caps do);
- More efficient lamps can be used (compact fluorescents now and solid state lighting in a few years);[4]
- Occupancy sensors that adjust lights and temperature to match need are available;
- Integrated system design should become the norm: for example, don't just ask if double glazed windows reduce the electricity bills, but ask also if you can use smaller and less costly heating and air-conditioning systems;
- More efficient electronics of all kinds needs to be introduced – standby power in all sorts of electronics (called vampires) consumes lots of unnecessary energy.

Most of these improvements have costs that are paid back in only a few years from the value of the energy saved. The management

---

[4] In commercial buildings lighting has been almost completely converted to fluorescents while in residences, there are many lights left to convert. One of the mysteries to be left to the social scientists is why the residential sector is so slow to change since compact fluorescents save much money. A 25 watt compact fluorescent lasts about 8000 hours while the equivalent in light output, a 100 watt incandescent, lasts only 800 hours. Over 8000 hours the incandescent will use about $80 of electricity while the fluorescent will use $20. In addition, over the same 8000 hours you will have to buy 10 incandescent bulbs at a cost of $19 compared with $5 for the one fluorescent. The total saving is $74, so why does anyone buy the incandescent light bulbs?

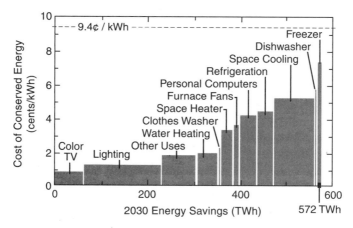

Fig. 11.8   Residential electric savings potential for year 2030. Conservation supply curve for electric energy-efficiency improvements in the residential sector. For each measure considered, the energy savings is achieved at a cost per kWh less than the average residential retail price of 9.4 cents per kWh, shown as the dashed horizontal line. (*Source*: Rich Brown, Sam Borgeson, Jon Koomey and Peter Biermayer, *U.S. Building-Sector Energy Efficiency Potential*, Lawrence Berkeley National Laboratory Report LBNL 1096E, September 2008)

consulting company McKinsey & Company released a report in 2007 that looked at the cost of reducing CO_2 emissions and found that many of the improvements in the building sector have negative costs in the sense that the savings outweigh the cost of the improvement [29]. Other studies have reached similar conclusions. Figure 11.8 is from a recent paper that looks specifically at the building sector [30].

Figure 11.8 is called a conservation supply curve and shows that all of the areas indicated have costs per kilowatt hour less than the average cost of the electricity and hence save money. The total electricity saving is about 2 Quads per year which translates to 6.6 Quads of primary energy, more than half of the required reduction in use to meet the goal of no increase in energy in the building sector from now to 2030.

The efficiency mystery

If installing all these improvements saves money and reduces emission, why are they not done in the United States? At the beginning of this chapter I said that in market economies, either consumer demand for energy efficiency forces manufacturers to comply with that demand in order to be competitive, or manufacturers see an advantage and work to convince the customers to buy what the manufacturers want to sell, or the government sees some national importance that makes it force efficiency on the society. Consumer pull works much better in the transportation sector than in the building sector. A typical driver travels about 12 000 miles per year, and the difference between a vehicle that gets 20 miles per gallon of fuel and one that gets 30 mpg amounts to about 200 gallons per year. When gas was $4.00 per gallon the cost difference was $800. It is even more significant to the economy as a whole. The total miles traveled in the United States by light vehicles in 2007 were about 3 trillion, and for that the difference between the 20 mpg and 30 mpg vehicle fleet amounts to 50 billion gallons of fuel. At $4.00 per gallon that amounts to $200 billion.

In buildings the situation is very different. It is difficult, sometimes impossible for the individual to find out what the savings might be from more efficient systems. Often the savings for the individual are small while the savings for society may be large in both cash and greenhouse gas emissions. Here is a sampling from the APS report [24] on barriers faced by consumers, manufacturers, builders, and designers of products to making them more energy efficient.

- **Not knowing:** You may know your total utility bill, but you do not know the contribution of each device that uses energy without some sort of required labeling system.
- **Not caring:** If the energy saving is small enough, the individual may not care about its cost. In 2002 TVs used a standby power of about 6 or 7 watts. Over a year that amounts to about $5.00 worth of electricity per TV, but over the 300 hundred million TVs it amounts to $1.5 billion and 10 million tonnes of greenhouse gases from our present mix of electricity sources. The new standard set by the federal government will be 1 watt.
- **Wrong incentives:** If the energy used in a building is not paid for by the building owner, the incentive is to install the lowest-cost systems rather than the most efficient systems.
- **Stalled innovation:** If manufacturers do not produce efficient products, consumers have no choice but to purchase what is available even if it is not very efficient.
- **Utility profits coupled to sales:** If utilities can make more profit only by selling more energy they have no incentive to promote efficiency. This is the situation in most states.

It seems clear that only regulation will drive the buildings sector to more efficiency.

What works

Appliance standards: In 1978 California set the first energy standards for refrigerators, furnaces, air conditioners, and other appliances sold in that state. Since the California market was so large, many manufacturers produced all of their products to meet the California standard. In 1987 nationwide federal standards were imposed that included more products, and appliance energy use continued to decline. For example, compared with 1972, refrigerators today use only 23% as much energy, gas furnaces 77% as much, and central air conditioners 60% as much. Many more appliances now fall under federal standards and the energy savings continue to increase as older systems wear out and are replaced with newer, more efficient ones.

The APS report estimates that by the year 2010 electricity use will be 7% below what it would have otherwise been, with an associated decrease in greenhouse gas emission of 240 million tonnes. By the year 2020 the electricity saving will be 11% and the greenhouse gas reduction 375 million tonnes. They also estimate that the savings to the economy are more than $300 billion more than the cost of implementing the standards.

Demand side management (DSM): DSM is really about reducing consumer demand through efficiency programs that are mainly run by the utilities. Here too, California has been a leader. Since the mid-1970s California's energy electricity consumption per person has remained constant while it has gone up by nearly 40% in the rest of the nation (New York has done as well). One of the key parts of the program that caused this to happen is called "decoupling" where utility profits are decoupled from sales. Before decoupling, regulated utilities could make only a fixed percentage of sales as profits. To make more money they had to sell more energy. After decoupling, they could make money by reducing demand for energy, and so began an era of free energy audits and other measures to get people to use less energy. Though this increased the cost of electricity per kilowatt-hour, the decrease in energy use more than made up for it and consumer expenses fell. About a quarter to a third of the energy savings during the period are estimated to come from this DSM.

Building energy efficiency standards: In the United States, energy codes are adopted at the state level. They are most often based on model codes developed by the International Code Council, or the American Society of Heating, Refrigeration, and Air-Conditioning Engineers. These codes are based on consensus, are not very strict, and when adopted are not rigorously enforced. There are a few exceptions in the United States, and once again it is California that is the leader. The California Energy Commission says that the energy savings from these codes has already amounted to about $2000 per household since they were first adopted in 1976. The residential code has been revised several times, most recently in 2008.

The best energy efficiency building codes are in Europe, where the energy use per square foot of floor space tends to be lower than in the United States. The strictest standards may be in Switzerland, though I confess I have not looked at the codes of many of the European Union members. There is a move in the United States to sharply reduce energy consumption in buildings over the next few decades. Among the states, California is once again taking the lead. Their new goal is that new residences use zero net energy (ZNE) by the year 2020, and new commercial buildings use ZNE by 2030. ZNE is defined as having enough clean electricity generated on site so that averaged over the year no net energy is used from the electrical grid. It is not clear what ZNE means with respect to heating.[5] The federal government has a goal for all new federal buildings of ZNE by 2030, and their definition is that electricity usage be reduced by 70% and the remaining 30% come from carbon-free energy generation either on site or off site.

The APS report analyzed the 70% energy reduction goal and concluded that for residences it is achievable by 2020 except for hot and humid sections of the country, but that achieving the commercial-building goal was going to need more advanced development. My conclusion is that for the last 30%, for a change the Federal approach is smarter than California's approach. The goal is greenhouse gas reduction and you should aim for that goal with the most cost-effective methods. The

---

[5] In Germany residences are already being built that use no furnaces for heating. They are designed with heavy insulation and are sealed against air leaks. Fresh air is brought in and the heat from the air being exhausted is used to heat the incoming air. Heat is supplied from the other appliances in the home and temperature is adjusted by varying the air intake.

Federal approach allows that, while the California approach mandates on-site energy generation. I live in the San Francisco Bay area and we are noted for our fog near the coast. If you live in the fog zone, you are not going to do well with solar power.

11.4 CONCLUSION

Transportation

The International Monetary Fund has projected that by the year 2050 there will be three billion cars on the world's roads compared with the mere 700 million on the road today. There is no way that today's vehicles can be used because there will not be that much oil to make gasoline. Something new is needed, and there are many things on the horizon that have the potential to allow that many vehicles to be run while reducing greenhouse gas emissions. It will be hard to get beyond the loudest and richest lobbyists to the most effective programs, but we have to try.

In the light-vehicle sector, improvement can come rapidly. The US CAFE standard goal of 35 mpg by the year 2020 will easily be met. I believe that the APS suggestion of a new standard of 50 mpg by the year 2030 for non-PHEVs is conservative. Europe is already much better than the United States. There are opportunities in advanced conventional hybrids, diesel engines, and new style gasoline engines. We should do it.

The PHEVs are likely to be revolutionary, and just how revolutionary will depend on the development of more advanced batteries. The first all-electric EV-1 came out in 1996 with conventional lead-acid batteries of the type used in all cars and had a 60 mile range. In 1999 the second generation EV-1 came out with advanced batteries, nickel-metal-hydride, and had a 120 mile range. In 2008 the Tesla has come out with lithium-ion batteries and a 240 mile range. In only 12 years the range of a small electric car has gone up by four times, and more advanced batteries are on the way. What is needed is more funding for long-range research to bring to reality batteries of a new type that are still more advanced. All too often funding agencies focus on near-term results which, though important, must not drive out the long-term efforts that are capable of revolutionary change. Battery development is advancing world wide. Japan, China, India, Europe, and the United States are investing heavily and there should be major progress in the next five or six years.

The alternative fuels program in the United States, is, to put it politely, misguided. I have not met a single scientist or engineer who believes that the current corn-based ethanol program makes any energy or environmental sense. It is mainly an agribusiness subsidy and they, of course, think it make lots of sense. There is potential in more advanced biofuels programs, but none of them have as yet proved practical. We should kill the mandate for more corn-based ethanol.

Hydrogen as a fuel for the light-vehicle fleet is questionable. The fuel cells themselves need to be sent back to the laboratory to emerge if possible with decent efficiency and with reasonable catalysts. Using natural gas on board a vehicle as a source of hydrogen may make sense, but I have yet to see an end-to-end analysis. The chain from electricity, to hydrogen, to pipeline, to vehicle, to electricity makes little sense. There are people working on biological sources of hydrogen and these may make sense.

Buildings

The normal way things are done in the business world is to do something if it saves money, but that does not seem to work in the buildings sector. If the objective is to decrease energy use and thereby decrease greenhouse gas emission external pressure has to be applied.

Appliance standards have already been shown to be cost-effective and we need more of them. The American Council on an Energy Efficient Economy (www.ACEEE.org) estimates that present appliance standards in the United States have already saved about 250 billion kWh of electricity which at a cost of 10 cents per kWh amounts to $25 billion. Since appliances continue to wear out and be replaced by newer models, the same standards are estimated to save another $230 billion in today's dollars (the ACEEE estimate is $160 billion in 1997 dollars) over the period from now to 2020.

The ACEEE gives a collection of suggested additions to the list of items having standards set and estimates that implementing those would give benefits amounting to about five times the costs of implementation. The problem in implementing standards for new items is the disconnect between those who bear the cost and those who reap the benefits. The manufacturer's cost goes up while the user's costs go down. A manufacturer is afraid that if he alone implements a money-saving item that increases his price while his competitors do not, he will lose business. The way around this is to have a requirement that all have to meet. We need more items covered by energy-efficiency standards and these standards need periodic updating as technology improves.

Demand side management (DSM) has been a winner in states that have implemented DSM programs. According to the Department of Energy's EIA, in the year 2006 DSM programs reduced electricity demand by more than 50 billion kWh at a saving of around $5 billion (there may be some double counting between the effect of appliance standards and DSM). The potential saving is much larger because only a few states have DSM programs. They were much easier to design in the days when all utilities were regulated, but that has changed with deregulations. Still, there are many states still regulated that have not adopted such programs. They can be done in deregulated states too, but are more complicated to design because the price of electricity from the generator is not controlled. I can only recommend to those wiser and more experienced in regulation than I that these programs be expanded.

Energy standards in the United States for buildings will be hard to implement without getting full cooperation from the states. It is hard to get Congress to take away something that the states think is theirs. Additionally, there are many climate zones in the country, and each needs a somewhat different standard. The simplest thing to do is to set the energy use per square foot for a building and let the building industry decide how to meet the standard in the most cost-effective way. The federal government will need to support additional R&D for buildings, and I hope that what comes out will do better than California in practice and as well as California in spirit.

The APS has suggested that all buildings should have an energy audit required before a residential building is sold to a new owner. I had to have a termite inspection made and given to a prospective purchaser before I sold my house. Adding an energy inspection would overcome one of the barriers to efficiency – not knowing. I would require the same thing for commercial buildings as well, though here it will mainly go to those leasing space. I don't think the real estate industry will like it but it might be a great help in promoting efficiency.

Efficiency is part of the solution to the problem of greenhouse gas emission, but it cannot do it all. The next several chapters will discuss energy sources that have little or no greenhouse gas emission. I will include nuclear, solar, wind, and geothermal systems which really do have little in the way of emission, and biofuels some of which pretend to have low emissions. After that I will discuss the policy issues that make devising a worldwide program very difficult.

Technical Note 11.1: CAFE standards

The CAFE standard is about the gasoline consumption of a fleet of different vehicles. It is not the average mpg of the fleet, but is related to the amount of fuel required to move the entire fleet by a given distance. It is easier to use an example than to give the equations which won't mean much to most people. Suppose you have a two-car fleet. One of them gets 10 mpg while he other gets 100 mpg. If you drove both for one mile the 10 mpg vehicle would use a tenth of a gallon of gas while the 100 mpg vehicle would use one-hundredth of a gallon. The total for the 2 miles is 0.11 gallon so this two-car fleet has a CAFE average of 2 divided by 0.11, or 18 mpg. If my fleet has two 10 mpg vehicles and three 100 mpg vehicles, the total fuel used in moving all of them one mile would be 0.23 gallons. Five total miles moved divided by 0.23 gallons of fuel used is about 22 mpg. I can use the same method if I have 10 different models in a corporate fleet with different numbers of each model on the road. It is even more complicated when different types of vehicles are held to different standards. That was the case before the new standards and is still the case, though today the different requirements are based on size rather than weight as they were before. For more details see the website of the National Highway Traffic Safety Administration (http://www. nhtsa.dot.gov/) and go to their section on fuel economy.

12

Nuclear energy

12.1 INTRODUCTION[1]

Nuclear energy is having a growth spurt. At the end of 2008 there were 435 nuclear power reactors operating in 30 countries, producing 16% of world electricity. Because of them, CO_2 emissions from electricity generation are three billion tonnes less than they would be without them (life-cycle emissions are shown in Figure 10.2). There were 28 new reactors under construction, mostly in Asia, and more than 200 more in the planning stage, including 30 in the United States.

Economic growth is driving demand for more energy, and concerns about energy supply and cost of fuel dominate the move to more nuclear power. The emission-free nature of the system is an environmental bonus. In all energy sectors of the world economy, demand for electrical energy is growing fastest (including for transportation), and how that electricity is made will determine how much and how fast greenhouse gas emissions can be reduced. Nuclear energy will play an important role everywhere, but perhaps not in the United States because of misplaced concerns about nuclear waste and radioactivity, and what may be the clumsiest system for making governmental technical decisions that could be devised.

First, a bit of history: when I was studying physics at MIT in the 1950s, nuclear physics was part of the standard curriculum. The nucleus and its constituents were then thought to be the smallest things (no longer so), and every physics student was expected to know the basics.

[1] A disclosure – I am a member of the DOE Nuclear Energy Advisory Committee and chair one of its subcommittees on advanced methods of treating nuclear waste. Until early 2008 I was on the Board of Directors of the US subsidiary of the French nuclear reactor builder, AREVA. I am no longer affiliated with any nuclear energy company.

We all knew the theory of how a nuclear reactor worked and even how a nuclear bomb worked. President Eisenhower's Atoms for Peace speech to the UN General Assembly in 1953 was exciting because it envisioned a world where nuclear weapons would be controlled and limitless nuclear energy would transform society. It didn't work out that way. The Cold War became our preoccupation and bombs rather than energy became the nuclear focus of the East–West rivalry.

Even so, nuclear power did advance in the United States and abroad until the accident at Three Mile Island (TMI) in Pennsylvania in 1979. Through a series of errors, the reactor cooling water was lost and the core melted. This was thought to be the most serious possible accident. Those in the United States who are old enough may remember the movie "The China Syndrome," with Jane Fonda. In the movie, which came out at about the same time as the TMI accident, a core meltdown occurred and the molten core was supposed to go on to melt its way through the bottom of the reactor, through the floor of the building and down into the Earth, causing terrible things to happen. I don't remember how the hero and heroine saved the situation, but save it they did.

In the real world, the TMI reactor core-meltdown caused no significant harm outside the reactor building because of safety standards that included a requirement for a containment building strong enough to hold any material that came from a damaged reactor vessel. However, it caused a rethinking of the operating systems on nuclear reactors, and required the modification of those already built and the redesign of those under construction. Delays dramatically increased costs and no new reactors were ordered in the United States after that time. Nuclear reactors continued to be built in many other parts of the world until the Chernobyl accident in 1986. This did cause much damage and some countries, particularly in Europe, stopped building new reactors, some planned to shut down their old ones, while others continued a nuclear power build-up.

Today there are 103 nuclear power plants operating in the United States without the emission of any greenhouse gases. Replacing one gigawatt of electricity generated from coal with the same amount generated from nuclear power would reduce CO_2 emissions by eight million tonnes per year. Nuclear electricity is available 24 hours a day, 7 days of the week (called base-load power). In the United States it supplies 20% of electricity; in Japan 30%; in South Korea 40%; and in France 80%. It is increasing from its worldwide base of 16% with many nations new to nuclear reactors showing interest.

France, with 80% of its electricity from nuclear reactors that emit no greenhouse gases, should be the poster child of the environmental movement. The country emits less than half the world average of greenhouse gas per unit GDP. If the entire world was like France, we would reduce carbon emissions by half, cutting them by about 3.5 billion tonnes per year (3.5 billion tonnes of carbon amounts to 13 billion tonnes of CO_2) and would have much more time to bring global warming under control. Yet the opposition to nuclear energy has been strong enough (mainly from countries in Western Europe with the exception of France) to prevent nuclear power from being accepted in the "Clean Development Mechanism" in the 1997 Kyoto Protocol, which gives extra credits to energy sources that emit no greenhouse gases. Opposition remains in the United States and parts of Europe, but seems to be weakening. Some prominent environmentalists have changed their minds about nuclear power because of concern about global warming.[2] (A primer on how nuclear reactors work is given in Technical Note 12.1 at the end of this chapter.)

The antinuclear movement argues that nuclear energy is dangerous because of radiation, reactor safety issues, and nuclear waste disposal; is expensive; and increases risks of weapons proliferation. These are serious issues, which will all be discussed in this chapter. Here is a quick preview.

- **Radiation:** We each get 10 000 times the radiation from naturally-occurring radioactive materials in our own bodies than we would get living next door to a nuclear plant. Radiation is dangerous and we have to control exposures carefully, but this is not an issue for a properly operating plant.
- **Safety:** Accidents are the issue, and a strong regulatory system is necessary as is proper design of the reactors. Design requirement differences are the reason Three Mile Island caused so little damage while Chernobyl caused so much.

[2] For example James Lovelock, leading environmentalist, creator of the Gaia theory, quoted in the British newspaper, *The Independent*, May 24, 2004; Patrick Moore, leading ecologist and environmentalist, one of the founders of Greenpeace, Chair and Chief Scientist of Greenspirit, quoted in *The Miami Herald*, January 30, 2005; Hugh Montefiore, former Bishop of Birmingham (UK) and former chairman and trustee for Friends of the Earth, quoted in the British newspaper *The Tablet*, October 23, 2004; Stewart Brand, noted environmentalist and founder, publisher, and editor of *The Whole Earth Catalog*, quoted in *Technology Review* (MIT), May 2005.

- **Waste disposal:** Spent fuel disposal seems to be a problem only in the United States. Other countries with large nuclear programs have approved plans. Some will bury the spent fuel. Others will bury part and save the most troublesome parts for use as fuel in the future.
- **Cost:** Nuclear electricity in the United States costs about the same as that from coal and is much cheaper than that from natural gas. France has the lowest-cost electricity in Western Europe.
- **Weapons:** Weapons proliferation is a serious problem and has to be controlled, but no proliferators have as yet gotten the necessary material from a civilian power reactor.

12.2 RADIATION

We all live in a continual bath of radiation. It comes from natural sources that are in our buildings, in the Earth, in the air, in our own bodies, and in the cosmic radiation that continually bombards us from space. We get still more from medical diagnostics like X-rays. We are born into this bath and live our lives in it. Our average life expectancy has steadily increased through the last century and is now nearly 80 years throughout the developed countries. It continues to increase while the natural radiation remains unchanged. Clearly, natural radioactivity has not imposed any limits as yet on life span. A sense of proportion is needed in thinking about radioactivity and radiation from power plants. The question should be, is radiation from nuclear power significant compared to the natural radiation we get all the time? It is not.

The unit of radiation is called the roentgen (designated R and named after Wilhelm Conrad Roentgen, the discoverer of X-rays). It is used to measure the total radiation dose received. It is a relatively big unit compared with natural or medical radiation, and people usually talk of milli-roentgens (mR), which are one-thousandth of a roentgen. Table 12.1 gives typical radiation doses from various sources. Natural and unavoidable radiation does vary from place to place, and the table gives the average dose received by a person at sea level in the United States from all the natural sources mentioned above.

The largest component of natural radioactivity is radon gas, which comes from the radioactive decay of the uranium and thorium that has always been part of the planet for the 4.5 billion years of its existence. It varies in intensity depending on location. The cosmic-ray

Table 12.1 *Typical yearly radiation doses*

Source	Radiation dose (mR/year)
Natural radiation	240
Natural in body*	29
Medical (average)	60
Nuclear plant (1 GW electric)	0.004
Coal plant (1 GW electric)	0.003

*Included in the natural total for a 75 kg person.
Source: National Council on Radiation Protection Report
160 (2009)

portion comes from particles that have been racing through space for longer than our planet has existed and continually bombard us. Its intensity depends on altitude, and at a high altitude, in the city of Denver for example, it is more than double that at sea level. The medical portion can be higher or lower, but the number used here is typical for the United States.

The second line in the table, "Natural in body", surprises most people. Our bodies contain traces of the elements potassium-40 and carbon-14. Potassium-40 is very long-lived and, like uranium and thorium, has been with us since the planet was formed. Carbon-14 is continually produced by the same cosmic rays that have been bombarding everything since the beginning of time. The radioactive forms of carbon and potassium are present in trace amounts in the food we eat and the air we breathe. Most of the "natural in body" radiation dose comes from the potassium in our bones. The annual dose listed in the table is for a 75 kilogram (165 pound) person: the dose is roughly proportional to body weight. If you sleep in a double bed you will get a few extra mR from your companion.

The comparison between natural radiation and a properly operating nuclear plant is striking. We get 50 000 times the radiation from natural sources than we would from living next to a nuclear plant and 10 000 times as much from material in our own bodies. Coal-fired power plants generate about as much radiation as nuclear plants from impurities in the coal. Radiation from a power plant is not significant compared with what we get all the time from natural sources, and we should stop worrying about it. The concern should be directed to reactor accidents and their consequences.

12.3 SAFETY

As mentioned above, there have been two serious reactor accidents, one in 1979 at Three Mile Island (TMI) in the United States, and one in 1986 at Chernobyl in the Ukraine (then part of the Soviet Union). The consequences were very different because of differences in the design of the reactors and the reactor buildings. All Western power reactors (almost all are light water reactors designated as LWRs – see Technical Note 12.1) are built with containment buildings that can hold radioactive materials inside in the event of an accident. The Russian RMBK reactors, the type used at Chernobyl, are built differently from LWRs, and have no containment building.

At TMI, when operations resumed after a regular maintenance shut-down, a series of start-up errors occurred, mainly related to the settings of the valves that controlled the flow of cooling water to the reactor core. Compounding the problem, the control system design did not give operators much information about exactly what was happening. The operators could not tell that certain valves that should have been open were closed and some that should have been closed were open. As operations began, the level of cooling water in the reactor vessel slowly dropped, exposing part of the core which then melted, releasing a large amount of highly radioactive material. It took 16 hours to bring the situation under control. The containment building kept almost all of the radioactive material inside, but some radioactive gases did escape. The subsequent investigation determined that the average exposure to the 2 million people in the region was 1 mR (less than 0.5% of natural radiation), while those close to the reactor (close is not well defined in the report) received exposures of about 10 mR. The maximum exposure to any one person totaled 100 mR (less than half of natural radiation).[3] These numbers are so small compared with years of exposure to natural radiation that there has been no measurable effect on the regional cancer rate.

The consequences of TMI to the nuclear power industry were profound. Many orders for new reactors were canceled and no firm new reactor order has been placed in the United States since, although 30 are now pending. Public attitudes toward nuclear energy shifted with approval ratings dropping to about 50% until recently, when they have

[3] The Nuclear Regulatory Commission's website has a summary of the TMI accident report. Go to http://www.nrc.gov/reading-rm/doc-collections/fact-sheets/3mile-isle.html to see it.

begun to be more favorable. More important for the future were the actions of the Nuclear Regulatory Commission (NRC). Design changes were mandated in all reactors then under construction and all operating reactors had to have retrofit programs approved. During this period of change, construction time for new nuclear plants stretched out to 10 years from the 5 to 6 years that had been the norm before. The cost of construction rose dramatically because of the delays, possibly spawning the notion that nuclear power is more expensive than others.

The period from 1979 to 1987 was a time of turmoil in the nuclear power industry as it argued with the NRC about what expensive fixes needed to be done. A changed occurred around 1987 when the nuclear power industry started to share more information among its members about operations and problems, and decided that cooperating with the regulators was better than arguing with them. This led to a remarkable improvement in the efficiency of nuclear power plants from a pre-1987 typical output of about 60% of capacity to today's 90%, the change coming from a sharp decrease in unplanned reactor shutdowns. Reactor operating hours increased 50% without spending any money building new reactors. The added electricity produced reduced the cost of nuclear electricity, and increased the profits of the nuclear-power industry. Today US power reactors are the best in the world in effective operation.

Memories of the Chernobyl disaster in 1986, the worst in the history of nuclear energy, still linger and should cause concern. More than 300 000 people were relocated because of radioactive contamination and the region around the plant is still full of ghost towns. The UN report on the accident estimated about 9000 excess cancer deaths over the lifetimes of the 6.8 million people who received significant exposures.[4]

The Chernobyl station consisted of four RBMK-1000 nuclear reactors, each capable of producing 1 gigawatt of electric power. The four together produced about 10% of Ukraine's electricity at the time of the accident. The complex was begun in 1970 and the one that failed, unit number 4, was commissioned in 1983. Reactors of the Chernobyl type have never been used for energy production outside the old Soviet bloc because of their potential to become unstable under certain conditions. In this type of reactor under unusual conditions the chain reaction can build up very fast, leaving no time for the reactor's

---

[4] The UN website http://chernobyl.undp.org/english/reference.html has links to many reports on the accident.

control rods to move in to stop an ultra-fast power excursion. Even for reactors of this type, the accident would not have happened had not the operators, for reasons that are still unclear, systematically disabled all of the reactor's safety systems. We know what they were doing because the control room log books survived. When the last safety system was disabled, the reactor ran away, blew off its own top along with the roof of the building, and spread radioactivity far and wide (the explosion was caused by superheated steam from the reactor's power build-up; it was not a nuclear explosion). The Chernobyl building was a light structure designed to keep out the weather rather than to keep in radioactivity from an accident. The European Union and the United States have cooperated in a program to have all such reactors modified to improve the safety systems or shut them down.

The new generation of LWRs now being built all over the world has been designed to be simpler to operate and maintain than the old generation. They have more passive safety systems, such as emergency cooling systems that rely on gravity feed rather than pumps that might fail to start in an emergency. Some designs claim to be safe in any kind of emergency without any operator action ("hands off" safe).

With a strong regulation and inspection system, the safety of nuclear systems can be assured. Without one, the risks grow. No industry can be trusted to regulate itself when the consequences of a failure extend beyond the bounds of damage to that industry alone. Recent examples of corrosion problems in a US reactor and in several Japanese reactors show again the need for rigorous inspections. Many countries that do not have nuclear power systems are beginning to think about implementing them. They will have to develop the technical talent to operate them properly, but it is at least as important that they develop the regulatory systems to keep them operating safely.

12.4 SPENT FUEL: LOVE IT OR HATE IT, WE HAVE IT

A political battle has been going on for over 20 years in the United States about what to do with the highly radioactive spent fuel that comes out of a reactor. After working on waste disposal for decades, there is still no licensed facility to store the 60 000 tons of spent nuclear fuel that have already come out of the power reactors now operating in the United States, much less the 120 000 tons that will come from them over their lifetimes. That doesn't count what will come from any new reactors that might be built. When I give talks on waste disposal I usually start with the title of this subsection: "Love it or hate it, we

Table 12.2 *Elements of spent fuel*

Component	Uranium	Fission fragments	Long-lived component
Percentage of total	95	4	1
Radioactivity	Negligible	Intense	Medium
Untreated required isolation time (years)	0	500	1 000 000

have it." It is hard for me to decide if the US waste disposal drama is a comedy or a tragedy. Here is the story.

US law gives the Federal Government responsibility to take title to all spent fuel and to put it away in a deep "geological repository," where it will remain isolated from the surface world for the time required for its radioactivity to decay to safe levels. That time is hundreds of thousands of years for the longest-lived component in untreated spent fuel. To pay for this repository, the price of electricity from nuclear reactors has a surcharge built into it of 0.1 cent per kilowatt-hour which goes to the government to pay for the eventual disposal of spent fuel. Over the lifetime of the reactors currently in operation, this waste disposal fund will accumulate about $50 billion, and there is about $20 billion in it now.

Looking separately at the three main elements of spent fuel (Table 12.2) might lead one to believe that there should be little problem. Uranium makes up the bulk and weight of the spent fuel. Nearly all of it is the uranium-238 isotope which has very low radioactivity and is not radioactive enough to be of concern. It could even be put back in the mines from which the original ore came.

There is no scientific or engineering difficulty in dealing with fission fragments (FF), the next most abundant component. Though very highly radioactive when they come out of a reactor, the vast majority of them have to be stored for only a few hundred years for radioactive decay to reduce the hazard to negligible levels. Robust containment that will last the required time is simple to build. The pyramids of Egypt have lasted more than 5000 years and there is little argument about our ability to do at least that well. There are two long-lived FFs, iodine-129 and technetium-99. They can be treated in the way the long-lived components are treated.

The problem comes from that last 1% of the spent fuel, composed of plutonium (Pu) and the elements called the minor actinides, neptunium (Np), americium (Am), and curium (Cm). The four are collectively known as the transuranics or TRU. Though they are much less radioactive than the FF, they are dangerous and have lifetimes 2000 times greater. Instead of isolation for hundreds of years, isolation for hundreds of thousands of years is needed. There is a second way to protect the public from this material, transmutation by neutron bombardment to change the TRU into shorter-lived fission fragments, but this is not yet out of the development stage.

Long-term isolation is the principle behind the "once-through" system, advocated by the United States from the late 1970s until recently as a weapons-proliferation control mechanism; the policy was adopted in 1977 by the Carter Administration. In once-through all the spent fuel is kept together. Plutonium, the stuff of nuclear bombs, in the spent fuel is not separated from the rest of the material, and so cannot be used in a weapon. Access to the plutonium is prevented by the intense radiation of the FFs that go with it into storage in a geological repository.

The once-through system may not be workable in a world with a greatly expanded nuclear-power program. The public wonders if the material can really remain isolated from the biosphere for hundreds of thousands of years. In addition, a large number of repositories would be required in a world with vastly expanded nuclear power. For example, even if nuclear energy in the United States were to remain at the projected 20% fraction of US electricity needs through the end of the century, the spent fuel in a once-through scenario would need nine repositories of the capacity limit set for our designated repository at Yucca Mountain in Nevada (the limit is by legislation; the physical capacity of Yucca Mountain is much larger). This would be quite a challenge since the United States has not yet been able to open its first one.

Yucca Mountain is right next to the Nevada test site, where hundreds of test nuclear explosions were carried out in the days of the cold war. There is a pretense that the existing radioactive contamination of the ground made it the logical site for a repository. However, the radioactive material from all the weapons tests ever made at the test site is about the same as that from two weeks of operation of the power reactors in the United States. The truth is that Nevada was chosen because it lacked the much larger political muscle of the alternate sites being considered.

In 1987, after a nationwide search for an appropriate site for the repository, three finalists emerged based on the geology of the available sites. One site was in Texas, the home of George H. W. Bush, then Vice President of the United States. The second was in the state of Washington, the home of Tom Foley, then Majority Leader of the House of Representatives. Nevada had little political clout at the time, and so, not surprisingly, was chosen. What was surprising was that Nevada was to get nothing for being the location of what Nevadans characterized as the nation's nuclear garbage dump. The only benefits Nevada would receive were the few hundred jobs that would go with operation of the repository.

The state has fought it ever since on any grounds it could, including that the siting procedures were flawed, the R&D was not done properly, the design was defective, and the radiation would not be contained. I think it is perfectly safe. Nevada now has one of its senators as the Majority Leader of the US Senate, so it is now politically strong where it was weak when the location of the repository was decided. The administration of President Obama has surrendered and announced that Yucca Mountain will not be used. Is this a comedy or a tragedy? Your call.

The alternative to once-through is a system based on reprocessing, which chemically separates the major components, treating each appropriately and eventually destroying most of the 1% of the spent fuel that produces the long-term risks. France's well-developed reprocessing system provides a good model that is described in Technical Note 12.2.

Until recently, the United States has opposed reprocessing on the grounds that it produces separated plutonium, which increases the risk that this material could find its way into nuclear weapons. In January 2006, President G. W. Bush announced a change in policy and initiated what is called the Global Nuclear Energy Partnership (GNEP). The purpose of GNEP is to develop a partnership of interested nations to advance the technology for continuous reprocessing and recycling the TRU in new fast-spectrum reactors that can in principle burn as fuel all the plutonium as well as the other long-lived minor actinides. When this is all worked out, the only materials that would go to a repository are fission fragments and a small amount of long-lived material that leaks into the fission-fragment waste stream in the separation process, which is not 100% perfect. The required isolation time would be roughly a few thousand years, a time for which isolation can be assured with very high confidence (remember the pyramids). It will take about 20 years to develop and test the technology.

We can temporize and continue to push the solution to the problem to later and later. All of the US operating nuclear plants have enough room to store on site all the spent fuel they will produce over their full lifetimes. I believe that this is also true for all or nearly all of the world's reactors.

Just as I do not like leaving the global warming problem to my grandchildren, I do not like leaving the nuclear waste problem to them either. It is solvable for either the once-through cycle or the separation-and-treatment cycle. My advice to our government is to store the spent fuel either at Yucca Mountain or at some other government site, or even at the reactors. Develop the technology for destroying the long-lived components in a new generation of reactors and if it all works out, go the reprocessing and treatment route and use a repository that needs to be secure for only 1000 years or so. If it does not, go the once-through and geological repository route. There is no real issue about Yucca Mountain's ability to contain material, but if the politics make it unusable, go somewhere else. There are other sites. Remember this about nuclear waste: love it or hate it, we will have at least 120 000 tons of it.

12.5 ECONOMICS

It is very difficult to compare the many estimates of the future cost of nuclear electricity. The estimates must include the cost of the plant, the cost of fuel, and the cost of ongoing operations and maintenance (O&M). Capital costs in some of the US studies reflect those of the period when it took many years to construct a plant and interest costs during construction piled up and up and up. Plant costs are usually given in dollars per kilowatt of electrical power output so that different plants can be compared on a common ground. The worst case was that of the ill-fated Shoreham power plant in New York that was never turned on because of local opposition, but ended up costing the builder $13 000 per kilowatt in construction costs including a large amount of interest on capital because of the delays in deciding the fate of the facility, saddling the region with one of the highest electricity rates in the country when the cost of this never-turned-on project were included in electricity charges.

In the period after TMI, costs of all reactors being built in the United States went up and it makes no sense to use those days as a predictor of the future. What does make sense is to use the numbers that go with plants recently completed or now under construction.

Many have been built in Asia and a few are under construction in Europe. In Asia six plants have been completed in Japan and Korea since 1994 at costs ranging from $2800 per kW for a plant completed in 1994 to $1,800 per kW for one completed in 2005 (2004 dollars). In Europe, a new plant is under construction in France and according to the French electrical utility Électricité de France (EDF) the cost will be about $4000 per kW (£3200 per kW), an increase from the originally expected cost said to be caused by increases in the cost of materials.

I have looked at many estimates of the future costs of nuclear power and by far the best is that of the World Nuclear Association [31]. There may be a natural suspicion that they as advocates would quote only the most favorable studies, but I have looked through their references and find them to be fair in their assessments. Others studies I have found to be flawed. For example, one recent one assumes a plant life of 40 years when the life is 60 years and may turn out to be even longer.

The important number is the cost of nuclear energy compared with other sources. The cost of fuel for a nuclear plant is about 35% of that for coal and about 20% of that for natural gas. The capital costs to construct a nuclear plant are larger than coal which is in turn larger than for natural gas. The cost of the plant is not all that has to be included; as important is the cost of money. Anyone who has ever had a mortgage on a home knows that interest costs are large and over the lifetime of the loan will probably be at least as much as the amount of the loan itself. Add in the volatility of concrete, steel, and other commodity costs and you have an economic guessing game, and you need to have a variety of guessers to have any hope of ending in the right range. Table 12.3 is from Ref. [31]. All but the European Union estimate are 3 to 4 years old. It is always difficult to correct an old estimate for conditions that change over the years. I would take the EU numbers as today's best estimate.

One of the options for nuclear energy for the long term involves reprocessing to get at the energy content of the plutonium in spent nuclear fuel. There have been questions about the cost of nuclear energy with this option compared with the cost using only fresh enriched fuel. A comprehensive analysis done at the Belfer Center at Harvard University [32] concludes that fuel costs with reprocessing are about 0.13 cents per kWh higher than those for fresh uranium fuel with a uranium price of about $40 per lb. This amounts to only a few percent premium in the cost of electricity and is small compared with the uncertainties in the prediction of future prices.

Table 12.3 *Estimates of electricity costs from various sources (US cent/kWh)*

	MIT 2003	France 2003	UK 2004	Chicago 2004	Canada 2004	EU 2007
Nuclear	4.2	3.7	4.6	4.2–4.6	5.0	5.4–7.4
Coal	4.2		5.2	3.5–4.1	4.5	4.7–6.1
Gas	5.8	5.8–10.1	5.9–9.8	5.5–7.0	7.2	4.6–6.1
Wind onshore			7.4			4.7–14.8
Wind offshore			11.0			8.2–20.2

First five gas-row figures corrected for January 2007 US gas prices of $6.5/ GJ (second figure for France and UK columns is using EU price of $12.15/GJ). Chicago nuclear figures corrected to $2000/kW capital cost. Canada nuclear shows figures for their new advanced reactor. Currency conversion is at 2007 values.

12.6 PROLIFERATION OF NUCLEAR WEAPONS[5]

Limiting the spread of nuclear weapons is a vital goal of the world community. There are three ways to get a nuclear weapon: build one, buy one, or steal one. So far, only the first method has been used. Since you can only buy or steal a weapon from states that already have them, the security of their weapons storage sites is the issue, not the expansion of nuclear-energy programs. The United States has spent lots of money helping the pieces of the former Soviet Union secure their weapons and weapons-grade materials, but that is another story, and buying or stealing an already-working weapon will not be discussed further.

Historically most of the nuclear-power industry has been concentrated in Europe, Japan, Russia, and the United States. Today, however, many new countries are planning reactors and making choices about their fuel supply that will determine the risk of proliferation for the next generation. The countries that traditionally set the tune for nuclear-power policies have waning influence on who goes nuclear, but they may be able to affect how they do it, and thus reduce the risk of weapons proliferation. The key is rethinking the "fuel cycle," the process by which nuclear fuel is supplied to reactors, recycled, and disposed of.

The design principles of nuclear weapons are known. While the technology required to make one is neither small-scale nor simple, it can be mastered by almost any nation. Examples of proliferators

---

[5] Parts of this section first appeared in an article by me titled "Reducing Proliferation Risk" in the Fall 2008 issue of a magazine of the National Academy of Sciences, *Issues in Science and Technology*, http://www.issues.org/25.1/richter.html

span a wide range of technological sophistication and include the very sophisticated (India, Israel), and the relatively unsophisticated (North Korea, Pakistan). The main obstacle to building a bomb is getting the fissionable material required.

The routes to obtaining the materials for uranium bombs and plutonium bombs are different. Enriched uranium comes from what is called the front end of the fuel cycle where raw uranium is enriched in the fissionable uranium-235. For a power reactor the enrichment target is about 4% to 5% (low enriched uranium or LEU) while for a weapon it is 90% (highly enriched uranium or HEU). The same process that produces the 4% material can be continued to produce the 90% material. This is what is behind the concern over Iran's plans to do its own enrichment.

Plutonium slowly builds up in the non-fissionable uranium-238 in the fuel whenever a nuclear reactor is operating. Weapons-grade plutonium comes from fresh fuel that has only been in a reactor for a few months and is nearly pure Pu-239, the favorite of the weapons builders. The plutonium from a power reactor where the fuel has been in for several years is called reactor-grade and has a mixture of several plutonium isotopes. It does not make as good a weapon, but the experts say that you can make one from reactor-grade material. To get at the plutonium, the spent fuel has to be reprocessed to extract the material from the radioactive spent fuel. Dealing with spent fuel is called the back end of the fuel cycle.

Clandestine weapons-development programs have already come from both ends of the fuel cycle (see Technical Note 12.3 on producing weapons material). South Africa, which voluntarily gave up its weapons in a program supervised by the International Atomic Energy Agency (IAEA), and Pakistan made their weapons from the front end of the fuel cycle. Libya was headed that way until it recently abandoned the attempt. There is uncertainty about Iran's intentions. India and Israel obtained their weapons material from the back end of the fuel cycle using heavy-water-moderated reactors, which do not require enriched uranium, to produce the necessary plutonium. North Korea used a related technology which also does not need enriched uranium.

There is no nuclear fuel cycle that can, on technical grounds alone, be made proliferation-proof to governments that are determined to siphon off materials for weapons. Opportunities exist for diversion of weapons-usable material at the front end of the fuel cycle where natural uranium is enriched to make reactor fuel. Opportunities also exist at the back end of the fuel cycle to extract fissile material from the spent fuel removed from reactors. While a completely diversion-proof system is impractical, one maxim can guide thinking on lowering the odds of proliferation: the more places this work is done, the harder it is to monitor.

The only way to prevent proliferation by a nation is through binding international agreements that include both incentives and sanctions. Close monitoring of the uranium enrichment process and of facilities where spent fuel can be reprocessed and plutonium extracted is also required so that there is an early warning when some nation strays. The IAEA headquartered in Vienna is the international organization that monitors all nuclear facilities. Its authority comes from the Non-Proliferation Treaty (NPT) that has been signed by almost all nations. The NPT gives all signers the right to develop nuclear power for peaceful purposes and the right to withdraw from the treaty if their national interest is threatened (Technical Note 12.4 gives the relevant sections of the treaty).

The two problem states that illustrate the difficulties of controlling proliferation even among those nations that have signed the NPT are Iran and North Korea. Iran has insisted that it will do its own enrichment as is allowed under Article IV of the NPT. Although Iran insists its program is peaceful, there is widespread suspicion that its real intention is to pursue a weapons program. The problem is that it is only a tiny step beyond enriching uranium for a power reactor to enriching some still more to make a weapon (Technical Note 12.3 gives numbers). Though sanctions have been imposed by the United Nations, they have not been effective.

North Korea built its own reactor and ran it for years. The small 5 MW reactor at Yongbyon was to be the prototype for a series of larger ones that were never built. Construction started in 1980 and operations began in 1985. North Korea signed the NPT in 1992 and a subsequent inspection of their spent fuel led the experts to conclude that by 1994, 30 to 40 kilograms of plutonium had been extracted from their spent fuel. When the North Koreans decided they needed weapons in 2002 they withdrew from the NPT invoking the "supreme national interest" clause (Article X-1) of the NPT, expelled the IAEA inspectors, and reprocessed more spent fuel from the reactor cooling pond. The response of the international community was ineffectual at first because there was no real agreement on what to do. Only in the past year or so have all North Korea's neighbors agreed on a response, and having China and Russia in the group pressuring North Korea has had some effect.

Both cases illustrate the problem. The science and technology community can only give the diplomats improved tools that may make monitoring the control agreements simpler and less intrusive. These technical safeguards are the heart of the systems used by the IAEA to identify proliferation efforts at the earliest possible stage. However, the technical safeguards cannot prevent attempts to circumvent the intentions of the NPT. Only an effective international response with agreed in advance sanctions can make the consequences serious enough to act as a real preventative measure.

There is no shortage of good ideas for creating a better-controlled global fuel cycle based on minimizing the number of fuel handling points. Mohamed ElBaradei, head of the IAEA, and former US President George W. Bush, for example, have suggested plans that would internationalize the fuel cycle. Enrichment and reprocessing would be done at only a few places and these would be under special controls. The problem is that such ideas, while good in theory, need to get the incentives for participation right. So far, these plans are the result of the "nuclear haves" talking among themselves and not talking to the "nuclear have-nots." While the talking proceeds, governments new to the nuclear game may conclude that they have to build their own fuel supply systems that are less dependent on suppliers with their own political agendas.

The problem needs urgent attention (Technical Note 12.5 discusses options for internationalization). If serious efforts do not begin soon the "have-nots" are likely to build their own fuel supply systems and the dangers of proliferation will become much higher. Plans to tame proliferation by nation states must include incentives to make any system attractive as well as effective monitoring and credible sanctions. Today, incentives are in short supply, inspections are not as rigorous as they could be, and there is no consensus on the rigor of sanctions that should be applied.

The scientists and engineers know that a major strengthening of the defenses against proliferation is a political issue, not a technical one. The politicians hope that some technical miracle will solve the problem so that they won't have to deal with political complications, but the scientists know that this is not going to happen. Short of a distant Utopia, the best that the nations of the world can do is to make it difficult to move from peaceful uses to weapons, to detect such activities as early as possible, and to apply appropriate sanctions when all else fails. Though there are technical improvements that can reduce proliferation risk, it is only in the political arena that real risk reduction can occur.

The real issue is the credibility of sanctions that can be imposed on those who violate the rules and start down the road leading toward a nuclear weapons program. There are few technical barriers to proliferators and if the international community does not act together no program can succeed.

If we continue on the business-as-usual course, global emissions are projected by 2050 to be about 15 billion tons per year producing 55 billion tons of CO_2, double that of today. A big piece of that comes from the generation of electricity where coal is now the king of fuels. Much of the electricity supply has to be what is called "base-load power," which is available all of the time. There is more to this than being able to watch your favorite television show at any hour of the day or night. Traffic lights, hospitals, airports, your home refrigerator, among many other things must all function 24 hours a day.

Nuclear power is the only large-scale carbon-free system that now can produce this base-load power. Expanding it to double the percentage of world electricity that it supplies today requires the deployment of about 1700 large new nuclear plants worldwide by 2050. With them, we could avoid the emission of about 3.5 billion tons of carbon per year. The United States has gotten stuck on the issue of safe disposal of radioactive waste. That is a political problem, not a technical one, but the political problem may delay the construction of the 30 nuclear plants that are now under discussion. If so, it will be much harder to reach the necessary greenhouse gas reduction target the new administration is discussing. In Europe outside France, opposition to nuclear power has been softening. A new reactor is under construction in Finland; the United Kingdom has said that nuclear power will be part of its energy future; Sweden is reconsidering its decision to phase out nuclear energy. In Europe it is only in Germany that the opposition to nuclear power remains firm.

A special word on the US role in nuclear power is in order. The United States was once the world leader in nuclear energy. It still has the largest number of power reactors (103) followed by France (59). Its reactors supply 20% of its electricity, but since the time of the Three Mile Island accident the US program has been in a systematic decline. It is no longer the leader in matters of policy because it has not been able to agree on one. It is no longer the leader in technology because bit by bit its R&D facilities in its national laboratories have been allowed to decay. It is no longer the leader in manufacturing because it is down to only one US-owned reactor builder, General Electric, and even that one works in partnership with the Japanese company Hitachi. It is the best in the world at operating reactors, having improved their uptime from 60% in the 1980s to more than 90% today, a remarkable feat that adds 50% to nuclear-electricity generation without building any new plants. The rest of the world does still have things to learn from the United States, but the United States has a lot to learn from them as well. The Bush administration made a start on a long-range nuclear plan, but it is an unfinished job and the new administration will have to finish it. Perhaps then the United States can join other countries in laying out a safe and secure nuclear energy future.

Nuclear energy is going to expand and countries that do not now have working nuclear reactors will want them. Such a future should include ways for countries new to nuclear energy to learn to operate and regulate their plants in a safe and secure manner, procedures to produce the needed fuel without increasing the risk of proliferation of nuclear weapons, and methods that all can use for disposal of radioactive material.

There is no one silver bullet to slay the climate-change dragon, and neither nuclear energy nor the renewables on their own can solve our greenhouse gas emission problem. What technologies might be available 50 years from now is beyond my vision. We need to get started and, for now, nuclear power provides us with one of the safest, most cost-effective alternatives to continuing on our present course. We should be moving vigorously to increase the nuclear energy supply.

Technical Note 12.1: Nuclear power primer

A nuclear power reactor is mainly characterized by three things: fuel, moderator, and coolant. The work horse of today's nuclear energy supply is the light water reactor (LWR). There are other kinds of reactors using different fuel, moderators, and coolants, but these are not widely used for energy production.

In an LWR, the fuel is enriched uranium (U). Natural uranium comes in two isotopic forms (the isotopic number is the total number of protons and neutrons in the nucleus), U-235 and U-238. Only the U-235 is fissionable and this makes up only 0.7%

of natural uranium. The LWRs of today use uranium enriched to about 4% to 5% of U-235, and enrichment today is done mainly with gas centrifuges. When a gas containing the two uranium isotopes is spun at high speed, the heavier isotope tends to move toward the outside. The gas near the center is taken off and goes to another centrifuge where it is further enriched. Each centrifuge gives only a tiny enrichment and a "cascade" of thousands of gas centrifuges is used to enrich the uranium to the necessary 4.5% level. The same process can be continued beyond the level required to produce reactor fuel to produce the 90% enrichment desired for nuclear weapons.

The moderator controls the energy of the neutrons from the fission process. When a neutron is captured in U-235, the nucleus will split (fission) into two lighter nuclei, releasing a large amount of energy plus a few more neutrons. The probability of neutron capture in a given nucleus depends on the energy of the neutron. Fission neutrons tend to have high energy, but the probability of capture and fission in U-235 is large for low-energy neutrons. The moderator, which in an LWR is ordinary water, controls the neutron energy. Neutrons collide with the proton nucleus of the hydrogen in water and lose energy at each collision, quickly reaching a low enough energy to make capture on U-235 highly probable.

The coolant in a reactor takes away the heat generated in the fission process, limiting the temperature rise in the reactor to what the design can stand without damage. It also transfers the heat to the power unit where the electricity is generated. In an LWR the coolant is ordinary water.

Another kind of reactor is moderated and cooled with what is called heavy water. Ordinary water, called light water in the reactor business, is composed of two atoms of hydrogen and one of oxygen. Heavy water has two atoms of deuterium, instead of hydrogen. The nucleus of deuterium has a neutron combined with the proton of ordinary hydrogen giving it twice the mass and hence the name heavy hydrogen. In the moderating process ordinary hydrogen can capture some of the neutrons while deuterium has a much smaller capture probability since it already has a neutron bound to the proton of the nucleus. LWRs need enriched uranium to make sure that a neutron sees a uranium

Technical Note 12.1 (*cont.*)

atom before it is captured. Heavy water reactors, because of the low capture probability in the moderator, can use natural uranium as fuel. They are not widely used because of their expense even after taking into account the savings from the use of natural uranium fuel. A relative of the heavy water moderated reactor is one moderated by carbon like North Korea's. It too can run on natural uranium.

Economies of scale have driven up the size of the present generation LWRs, most of which are in the gigawatt range. There is increasing interest in smaller reactors for places that are not connected to a high-capacity power grid or where the economy is not developed enough to be able to use the full output of a big reactor. Several groups have begun development of smaller reactors. Toshiba, the Japanese reactor builder, has developed its 4S reactor with an output ranging from 25 MW electricity (MWe) to 100 MWe. This design has a sealed core that only needs refueling every 30 years, making it much easier to operate and monitor for potential proliferation problems. Two other companies, Hyperion Power Generating Systems in New Mexico and NuScale Power in Oregon, are developing reactors in the 25 to 50 MWe class. Others are showing interest as well. The hope is that these will be small enough to be entirely fabricated at a central factory and delivered complete to the site, thus greatly reducing costs.

There are two good websites with information on all topics related to nuclear energy. They are those of the World Nuclear Association (www.world-nuclear.org) and the Nuclear Energy Agency (www.nea.fr). The NEA has a particularly good general overview called "Nuclear Energy Today."

Technical Note 12.2: France's long-range nuclear development plan[6]

France has what I see as the world's most well-thought-out and coherent long-range planning that lays out where they see their

[6] This section is based on an article of mine first published in a different form in *Newsweek International*, July 7–14, 2008.

nuclear program heading over the next 50 years. Their vision is built around a future that includes secure disposal of nuclear waste, advanced reactor development, and considerations of possible fuel shortages in the future. I look at it as a kind of 50-year-long superhighway with various on- and off-ramps that allow flexibility to handle changing technical options.

The highway was begun in the 1990s with the assumption that nuclear power will remain the mainstay of France's electrical generating system for the long term. In addition, they see nuclear power as having other potential applications that may expand nuclear power demand even faster than would occur from growth in the electrical system alone (process heat, hydrogen production, desalinization, etc.). Also, they assume a world nuclear power expansion, greatly increasing the demand for new reactors and reactor fuels. Today's LWRs are expected to remain the work horses of nuclear energy until at least mid-century when something new may be needed.

The highway starts with an LWR fleet initially fueled with enriched uranium. When a fuel load is used up it contains a significant amount of plutonium and that plutonium is extracted (reprocessed) and mixed with unenriched uranium to make a new fuel called MOX (a mixture of uranium and plutonium oxides). This new fuel is used to get about a third more nuclear energy from the original enriched uranium fuel than would otherwise be possible. The leftovers from the plutonium extraction process contain americium, neptunium, and curium which have to be isolated from the environment for hundreds of thousands of years.

An off-ramp for spent fuel involves the development of geological repositories that are safe for the very long term. Materials put into these repositories initially will be emplaced in a manner that allows them to be retrieved if so desired. If new technology comes along that can treat this material to make it easier to store, that new technology can be used, otherwise the repository will be sealed.

With the assumptions made on the expansion of nuclear power, a shortage of natural uranium needed for lifetime fueling of the LWR fleet may occur about mid-century. An on-ramp if new fuel is needed requires the development of a new generation of

Technical Note 12.2 (*cont.*)

advanced nuclear reactors that can breed new fuel from depleted uranium. The necessary R&D will be done over the next 30 years so they will be ready for large-scale commercial development around mid-century if needed and economical.

The same technology that can be used to breed new fuel can be tweaked to allow these reactors to destroy the long-lived components of spent reactor fuel, creating a second off-ramp for spent fuel. These new reactors can be used, even if not needed for power, to reduce the required isolation time for spent fuel from hundreds of thousands of years to around one thousand years. This new technology will be used to treat spent fuel from all reactors when the new technology is available.

R&D on new ideas and technologies will proceed in parallel with the main highway so that new on- or off-ramps can be constructed if needed.

The French long-range nuclear-energy plan was developed with the involvement of their electric utility, the company that builds their nuclear reactors, their CEA (the equivalent of the US Department of Energy), and their Parliament. As in the United States, the nuclear waste issue was the most contentious, and the contrast between how they handled it and how the US is still struggling with it is striking.

France has what is called the Parliamentary Office for Scientific and Technological Assessment (POSTA). It is a joint committee of their two houses of parliament, has a membership proportional to representation of the political parties in parliament, has a civil servant staff, and has a high-level external scientific advisory committee. The United States had something like it, too, but it became mired in political infighting and Congress abolished it in 1995 as part of the Gingrich revolution in the House of Representatives.

France passed a law in 1991 on the advice of the POSTA giving the government 15 years to report back with their proposal for handling nuclear waste. In 2005 POSTA began a series of hearings on the government's proposal (I testified at one). The result was the Act of 2006 blessing the road outlined above. In contrast, the United States has no coherent long-term policy though the administration is trying to craft one, and we have not been able to get our own repository started in spite of 20 years of effort.

Technical Note 12.3:   Producing material for weapons

Uranium: All LWRs use enriched uranium where the amount of the fissionable isotope U-235 is increased from the naturally occurring fraction of 0.71% in uranium ore to 4% to 5% depending on the design of the reactor. This enrichment process (the front end of the fuel cycle) is of concern because of the potential to enrich far beyond the requirement for power production to the level needed to make a nuclear weapon. Although anything enriched to greater than 20% U-235 is considered weaponizable, in reality any state or group moving toward a weapon will want material enrichment to the 90% level. It is much easier to make a weapon at this enrichment than at 20% enrichment.

If a facility is doing the enrichment required for a power plant, it takes only a small increment in capacity to produce the material for a few uranium weapons. Some numbers are useful to understand the problem. Most of the nuclear plants being built now use 4.5% enriched uranium. A 1 GWe nuclear plant requires about 20 000 kg of new enriched fuel per year. Because natural uranium contains much less than 4.5% of the U-235 isotope, nearly ten times as much natural uranium is required to make that much enriched fuel. However, only about 600 kg of the 4.5% enriched material is needed as input to make the 90% enriched material for a single weapon.

As described earlier, the preferred technology for enrichment now is the gas centrifuge. A cylinder of gas (uranium hexafluoride) containing both U-235 and U-238 is spun at very high speed. The heavier U-238 tends to concentrate more at the outside of the cylinder so gas taken off from near the center is slightly enriched in U-235. Since the enrichment is slight at each stage, a multistage cascade of centrifuges is needed to enrich to the level needed for reactor fuel.

While these devices sound simple in principle, the reality is quite different and the technology of the modern high-throughput centrifuge is not simple to master. Those used by Pakistan for their uranium weapons are primitive by the standards of today. Making enough fuel for the Iranian Bushehr reactor would need about 100 000 of the Pakistan P1 centrifuges (the design Iran started with), requiring a very large plant. It only takes about 1500 more centrifuges running on the output of the fuel fabrication plant to

Technical Note 12.3 (*cont.*)

make enough 90% enriched material for one uranium weapon per year. An entirely secret plant running on natural uranium would need about 5000 of the P1 to make enough 90% enriched material for one bomb per year. The more advanced Iranian IR-2 centrifuge announced in 2008 is said to have a throughput of about three times that of the P1, so all the above numbers could be reduced threefold.

The most advanced centrifuges are those of the British–Dutch–German consortium URENCO whose T12 units are said to have a throughput 30 times that of the P1 and whose next-generation T21 units are designed to have a capacity of 70 times that of the P1. A 1 GWe reactor only needs a plant with about 4000 T12 units to supply it with the necessary enriched fuel and with the T21 design would need only about 1500 units. To prevent illicit diversion or over enrichment, all the piping and connections among the centrifuges need monitoring. It is much easier to reliably monitor a plant with only a few thousand units than it is to monitor one with a hundred thousand units. On the other hand, it is much easier to hide a clandestine plant with only a few centrifuges. The advanced technology machines need very tight controls.

Plutonium: A 1 GWe LWR produces about 20 tonnes per year of spent fuel. Roughly, this material consists of about 1% Pu, 4% fission fragments, and 95% uranium. When spent fuel comes out of a reactor, the radiation is so intense that some form of cooling is required to keep the fuel rods from damage that would lead to the escape of radioactive material. They are put into cooling ponds where enough water circulates by natural convection to keep the temperature of the rods at a safe level. Typically, practice is to keep the fuel in the ponds for at least 4 years. The radioactivity and the associated heat generated have, by then, decayed enough to allow the rods to be removed from the ponds and put into dry casks for storage or shipped off site if so desired. If removed from the ponds, containers have thick enough shielding that the material can be safely stored above ground with no danger from radiation. The intense radiation from the fission fragments in the spent fuel acts as a barrier to theft or diversion. Since any would-be thief would receive a disabling and lethal radiation dose in a matter of minutes, the spent fuel including its plutonium is thought to be safe.

The spent fuel can be reprocessed to extract the plutonium and reuse it to generate more energy in their power reactors. The chemical process used is called the PUREX process and is well-known chemistry; there are no secrets to learn. This is the process that North Korea used to extract the plutonium for the spent fuel from their Yongbyon reactor to get material for their weapons program. There are really two periods in the life cycle of spent fuel with different proliferation vulnerabilities. The first period is while it is in the cooling pond at each reactor. The second period is after the radioactivity of the spent fuel has decayed to the point where it can be removed from the cooling pond. I separate things this way because there is no way as yet to move spent fuel from a cooling pond until its radioactivity has decayed enough that it can be moved with passive cooling, so there is a period when the spent fuel with its plutonium remains under the control of its place of origin. North Korea expelled the IAEA inspectors, reprocessed the spent fuel from its Yongbyon reactor and produced the plutonium needed for its first weapons. This is called in the trade a "breakout scenario". Everything is going according to the international rules, until suddenly they are not. A 1 GWe reactor produces about 200 kg of Pu each year so the 4-year inventory in the pond is enough to build roughly 100 plutonium weapons.

Technical Note 12.4: Extract from the Treaty on the Non-Proliferation of Nuclear Weapons

Article IV

1.   Nothing in this Treaty shall be interpreted as affecting the inalienable right of all the Parties to the Treaty to develop research, production and use of nuclear energy for peaceful purposes without discrimination and in conformity with articles I and II of this Treaty.

2.   All the Parties to the Treaty undertake to facilitate, and have the right to participate in, the fullest possible exchange of equipment, materials and scientific and technological information for the peaceful uses of nuclear energy. Parties to the Treaty in a position to do so shall also cooperate in contributing alone or together

Technical Note 12.4 (*cont.*)

pwith other States or international organizations to the further development of the applications of nuclear energy for peaceful purposes, especially in the territories of non-nuclear-weapon States Party to the Treaty, with due consideration for the needs of the developing areas of the world.

Article X

1.   Each Party shall in exercising its national sovereignty have the right to withdraw from the Treaty if it decides that extraordinary events, related to the subject matter of this Treaty, have jeopardized the supreme interests of its country. It shall give notice of such withdrawal to all other Parties to the Treaty and to the United Nations Security Council three months in advance. Such notice shall include a statement of the extraordinary events it regards as having jeopardized its supreme interests.

Technical Note 12.5: Issues in internationalizing the fuel cycle

The main focus has been on creating an attractive alternative to national programs. The exemplar is South Korea which gets 39% of its electricity from nuclear power plants and by its own choice does no enrichment of its own. It has saved a great deal of money because it did not have to develop the enrichment technology, nor did it have to build and maintain the enrichment facilities. To make this kind of proposal acceptable to those countries that are not firm allies of those that have enrichment facilities requires some mechanism to guarantee the fuel supply. Without such a mechanism, it is doubtful that any sensible country interested in developing nuclear power would agree to a binding commitment to forgo its own enrichment capability. The tough cases are not the South Koreas of the world, but such states as Malaysia, Indonesia, Brazil (which has two reactors and talks of building more), and many other states with growing economies that are more suspicious of their potential suppliers.

This is not going to be easy. If the supplier of enrichment services is another country, how could the first be sure that the second would not cut it off from its needed fuel for political reasons? Europe, for example, gets a large fraction of its natural gas supply from Russia through a pipeline that runs through Ukraine. In 2006 and again in 2009, in a dispute with Ukraine, Russia turned off the gas. It only lasted a short time, but Ukraine had to agree to Russia's terms and Europe's confidence in the reliability of its supply was badly shaken.

In the energy area, being very heavily dependent on a single source of supply is economically and politically dangerous. We have been through this with oil supply in the 1970s. The Arab members of OPEC cut off oil supplies to the West because of its support for Israel. It was disruptive, but we did get through. The US response was to diversify suppliers and to build reserve storage capacity.

Countries with a new or relatively small nuclear program will strongly favor international systems if they come to trust the suppliers of the fuel and other needed services. Today the only places to purchase enrichment services are the United States, Western Europe, and Russia. This group is too narrow in its political interests to be a credible system for supply. Others must be encouraged to enter the fuel supply business. A well-managed system in China would add considerably to political diversity in the supply chain. A reserve fuel bank under the auspices of the IAEA would also help.

Reducing the proliferation risk from the back end of the fuel cycle will be at least as complex as from the front end, but doing so is essential. North Korea has demonstrated how quickly a country can "break out" from an international agreement and develop weapons if the material is available. Here the thinking is about a limited number of international reprocessing facilities that would do the needed work for all comers. The consequences of denial of services to a subscribing country are insignificant. All that is required for continued operation is to refuel a reactor with fresh enriched fuel rather than with the output of a reprocessing facility.

13

Renewables

13.1 13.1 INTRODUCTION

Discussion of renewable sources of energy is where you will find the largest collection of half-truths and exaggerations. The Renewables covered in this chapter include wind-, solar-, geothermal-, hydro-, ocean-, and biomass-energy systems (biofuels are treated in the next chapter). According to the *OECD Factbook* (2008), renewables made up about 13% of world total primary energy supply (TPES), but the only two that make a significant contribution to emission-free energy today are large-scale hydroelectric dams and biomass. Large hydropower systems supply 18% of world electricity and 4% of TPES, but are often not included in the definition of renewables for reasons that involve value judgments that have nothing to do with greenhouse gases and global warming. Biomass, which contributes 7% of TPES, is the use of waste plant and forest materials for energy and is the fuel that the poorest people have available for heat and cooking as well as supplemental fuel for energy in more developed nations; I will come back to it briefly in the chapter on biofuels..

When large hydro and biomass are excluded, only a tiny part of TPES comes from wind, solar electrical, geothermal, ocean, and biofuel systems: less than 1% in the Unites States and less than 2% worldwide.[1] Of these, wind is the largest, supplying about 1% of US electricity in 2008, but has problems because it is intermittent and the best sites are often not where the largest demand is. Solar energy's problem is

---

[1] Three good sources for more in-depth information about renewable energy are the *OECD Factbook* [39], the report of the REN 21 group [35], and the US Energy Information Administration website http://www.eia.doe.gov/fuelrenewable.html. There are some differences in definition relating to primary energy, so read the fine print.

that the sun does not shine all the time and no good method of storing electricity exists. A new source of geothermal energy from deep, hot, dry rock is being developed which, if successful, will allow a big expansion of geothermal power, but it was tried in the 1970s and failed – the jury is still out on this one. The oceans are a harsh environment and ocean systems have not worked so far. New technology is being tried. Biofuels get a chapter of their own following this one. As of today they probably do more harm than good, except perhaps in Brazil. The potential of the next generation of biofuels is uncertain because none of them are ready for commercial-scale deployment and until they are no one can evaluate their total impact.

Here, I will review each including what I see as the associated issues that have to be addressed if the renewables are to reach their potential. There is much that can be done with present technology to increase their role, but there are serious problems with costs. A phrase that is used in the business of electricity generation is "grid parity": how the renewables compete in cost with electricity from the power distribution grid that is generated from the major sources, coal gas, nuclear, and large-scale hydropower. Today, none can compete without large federal and state subsides, and even with them only wind power comes close to grid parity. The intent of the subsidies is to stimulate large-scale development so the costs will come down with experience. They are coming down in some cases, but more slowly than initially expected. This will be discussed further in Chapter 15.

13.2 WIND

Wind power seems to have come from Persia sometime between 500 and 1000 AD.[2] Wind power was first used for grain milling, hence the name windmill. In Europe it was used for grain and for pumping water for irrigation, to dewater mines, and to pump water out of the lowlands of the Netherlands. Anyone who has traveled through the farm country of the United States has seen many small, multi-bladed windmills that are still used to pump water for irrigation and for watering livestock. Windmills were an important technology that came to a first maturity in the nineteenth century, and those are the ancestors of today's giant wind turbines that are now being used for a new purpose, generating electricity.

---

[2] A good short history of wind energy on the web is *Illustrated History of Wind Power Development* by D. M. Dodge at www.teleosnet.com/wind/index.html

The economic attraction of wind power is that its fuel is free. The environmental attraction is that no greenhouse gases are emitted (some environmental groups have concerns about injuries to birds). The giant wind turbines of today are expensive, and are not economic winners on their own compared with coal-generated electricity as long as coal plants do not have to account economically for their emissions. Even so, thanks to generous tax subsidies wind energy is profitable. It is not clear what role wind would play in the United States and elsewhere without tax breaks, but as things are, world wind power is growing by about 20% per year worldwide and faster in the United States. If emission charges were included in the cost of electricity, wind would probably be a winner without any subsidies.

There are good sites and bad sites for wind turbines. The good have fairly steady wind with at least moderate speed. The bad sites have highly variable speeds or steady low speeds. The electric power generated by wind grows rapidly with wind speed; doubling the wind speed increases the output of a wind turbine by eight times. Double the wind speed again and the power goes up by another eight times. That cannot go on forever, and there is a maximum speed for a wind turbine above which it has to be shut down to prevent damage. There is also a minimum wind speed needed to overcome losses in the system, so there is a band of wind speeds in which the turbine can operate effectively.

The ideal site for a wind turbine would have a steady wind speed at the turbine's sweet spot for energy output. However, nature is not so kind as to provide such sites, and a major problem for wind is its variable nature. Sometimes it blows hard, sometimes not so hard, and sometimes not at all. To understand wind's real energy contribution, two different numbers have to be looked at; one is the capacity of the wind turbine and the other is its output. The capacity is what is written on the turbine's name plate and says what the maximum output is. When you read or hear about a wind farm with a 100 megawatt capacity, you are reading about its maximum output in perfect conditions. The real world is very different, and no wind farm puts out its rated capacity for very long or even for a majority of the time.

Figure 13.1 shows the energy actually delivered by all the wind turbines of the German utility E.ON Energie during the year 2007 as a fraction of the total electrical energy produced by that utility. The fraction varies wildly mainly because of the variable wind speed. The wind turbines are spread over a large part of the German land area so that when the wind is not blowing in one area it may be

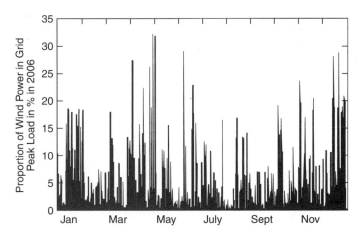

Fig. 13.1   Wind variability in Germany. Percentage of electrical demand delivered by all the wind turbines of E.ON Energie during the year 2007. Averaged over the year, wind power delivered 18% of installed capacity. (Courtesy of E.ON Energie)

blowing at another, but it is very rare for the wind to blow at all sites hard enough to give 100% of capacity as the actual output. The real year-long average output for this company's wind turbines is 18% of capacity.

The International Energy Agency (IEA) in Paris is home to a 20-country consortium called IEA Wind that collects data and exchanges information and experience with the aim of facilitating and accelerating the deployment of wind energy systems.[3] The members of the group have about 80% of total worldwide installed wind capacity. There are better and worse areas for wind energy than Germany. Table 13.1 shows the average wind output from several countries as a fraction of installed capacity; the E.ON result is not that unusual. The total installed world wind capacity is 94 GW, and at 18% efficiency the world wind turbine output is about equal to 1% of electricity consumption. Denmark is the wind champion in the fraction of total electricity generation from wind, getting 20% of its electricity from wind turbines. Though small today, installed capacity worldwide is growing by about 20% per year, giving a doubling time of three to four years. The largest increase in a single country was 46% in the United States in 2007, followed by an increase of 50% in 2008.

[3] See www.ieawind.org/annual_report.html for a summary of data from around the world.

Table 13.1 *Wind energy output as percentage of capacity (2007)*

Country	Wind capacity (gigawatts)	Wind output (terawatt-hours)	Average output (% of capacity)
US	17	48	32%
Denmark	3	7	20%
Germany	22	39	20%
Spain	15	27	20%
China	6	4	17%
India	8	8	12%
World	94	152	18%

The European Union has a target of 20% of its energy from renewables by the year 2020. Specific targets are yet to be set for each country in the EU. These are necessary because of differing potentials for renewables in different countries, and wind power plays a large role in their plans for the electricity sector. Whatever the details, the EU is committed to a big expansion in renewables and they will use all the renewables discussed in this chapter as part of the package.

I noted earlier that good wind sites are not uniformly distributed. Figure 13.2 from the DOE's Energy Efficiency and Renewable Energy Division (EERE) shows the situation in the United States – the darker the shading, the better the site.[4] The best are at the coasts, on the Great Lakes, and in spots on the Great Plains. Offshore is generally better than onshore because winds are steadier and speeds tend to be higher. However, offshore installations are more costly to construct and maintain than onshore, and there has been little offshore installation as yet anywhere in the world.

Figure 13.2 also illustrates another issue that has to be addressed to greatly expand the use of wind energy. The electric power distribution grid needs to be modernized to move power from the Great Plains region to the east and south to get it to where the large demand is. This, as we will see, is also an issue with solar electricity, and grid modernization is a high priority if wind and solar energy are to meet their expected potential. Although the map is for the United States, similar situations exist in all parts of the world. In the UK, the wind farms are in the north while the largest loads are in the south. Unfortunately,

---

[4] www.windpoweringamerica.gov/wind_maps.asp

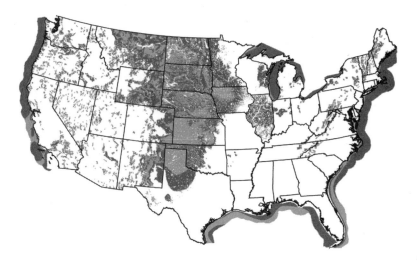

Fig. 13.2   United States – wind resource map. Wind quality for electricity generation. Darker is better. (*Source*: US Department of Energy, Energy Efficiency and Renewable Energy Division)

in the United States the only thing that takes longer than gaining the approval for a new nuclear power plant is gaining the approval for a new high-voltage electrical transmission line. It is to be hoped that things are better elsewhere, but high-voltage transmission lines are not beautiful and seem to excite considerable local opposition.

The variability of wind creates another problem: the need for backup electricity supply for when the wind is not blowing. When wind makes only a small contribution to supply, this is not a serious problem. Only a small standby system (usually natural gas powered) is needed to maintain output and smooth variations. As wind makes up a larger portion of supply this problem becomes larger too, and I am not comfortable with the analyses that I have seen because they all seem to be incomplete. The most recent is in a report from the EERE division of the US DOE entitled *20% Wind Energy by 2030* (www.eere.gov /windandhydro). It includes one assumption that is wrong and is missing a discussion of what I call correlations between wind farms in the same area.

The wrong assumption is about the independence of load variations and wind. In California, for example, we have the infamous Santa Ana wind: a hot dry high-speed wind that sometimes blows from the interior toward the coast. It is the cause of the worst damage from forest fires. But, also, when it blows demand for air conditioning goes up and the winds are so high that wind turbines have to shut down. To be sure, this increase in demand coupled to a decrease in supply is a rare occurrence, but rare conditions can cause a great deal of damage as we have seen in the recent economic meltdown and credit squeeze. In his recent book [33], N. N. Taleb called rare and important events "black swans". There are potential black swans in the energy area too.

The second issue is the correlation between the outputs of wind farms in the same region. This may have been analyzed, but if so I have not found it. Clearly, if the wind is not blowing in one place, it is not likely to be blowing only a few miles away. I asked one of the California experts about this and his answer was that the correlation length (the distance that you have to go to get very different conditions) was about 100 miles. That would say if the wind is not blowing at one wind turbine, it is not likely to be blowing at any other within a radius of 100 miles. Mathematically this is not a hard problem to deal with, but to do so you need what the mathematicians call the two-point correlation functions. I hope that they exist, but my California expert did not know of one for California. If they don't exist today they are needed to assess the real potential of wind energy.

The bottom line on wind as an energy source is that it is small now and growing. It should be encouraged as one of the systems to generate electricity with no greenhouse gas emissions. Wind now supplies about 1% of US electricity and can grow to 5% to 10% with no real problems. Above that level, to become a major component of the nation's or the world's energy supply, the power grid needs upgrading and some important questions about variability and backup need answering. There is considerable exaggeration about promises such as 20% wind by 2030 as given in the DOE report. With the output factor of 32% in the United States that would require an installed capacity of 60% of the total electricity supply, which is highly unlikely.

13.3 SOLAR ENERGY

There are really three solar-energy systems. The two that get the most attention are solar photovoltaic and solar thermal electrical generating systems. The American comedian Rodney Dangerfield used to complain in his routine that "I don't get no respect." The biggest solar energy contributor to energy and greenhouse gas reduction is the solar hot water and heat system and "it don't get no respect," though

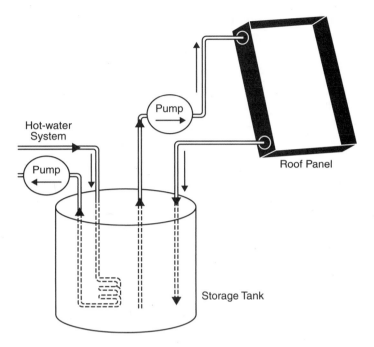

Fig. 13.3   Solar hot water. The system circulates water between a
storage tank and a roof panel heated by the Sun. A heat exchanger in
the storage tank produces hot water for the residence.

worldwide it represents the equivalent of 105 gigawatts of power compared with only 7.7 gigawatts for solar electricity.

Figure 13.3 is a sketch of the basic heat and hot water system.
Water is pumped to rooftop panels where it is heated by the Sun
and returns hot, heating fluid in a large storage tank (these systems
work in cold climates too, but have to have well insulated pipes and
antifreeze in the fluid). Domestic water is pumped through a heat
exchange coil inside the tank, picks up heat from the hot water in
the tank and supplies domestic hot water and in some cases heat to
the building. In the mid-1980s I installed such a system at my home
which had two heat exchanger coils. One supplied most of our domestic hot water, while the other supplied about half the heat needed for
the house. The insulated tank held 800 gallons of water and so could
supply heat for two to three days if needed. The system functioned
reliably for 20 years.

Until a few years ago Israel was the only country that required
such systems. Recently Spain, India, Korea, China, and Germany have
required that at least part of domestic heat and hot water come from

solar systems. Although the United States does not require such systems, it does extend the renewable energy tax credit to them. California has the most aggressive "renewable portfolio standard" in the States but only seems to include electricity generation from renewables (excluding large hydropower systems). Excluding hot water and heat from a solar system from the renewable portfolio seems silly since they also reduce greenhouse gas emissions by substituting the Sun for natural gas or electricity.[5]

Photovoltaic (PV) systems to supply electricity are becoming more popular. The silicon solar cell was developed in the 1950s. The first-generation solar cells were very expensive, and their first real application was in space systems, a very demanding environment where cost was of little importance because they were only a tiny fraction of the cost of a space project. Development has been continuous since the early days and there are now many kinds of cells of varying cost and efficiency. It is unfortunately true that the cheaper the cell the lower the conversion efficiency from sunlight to electricity, but even so the cost of electricity from solar cells has come down greatly from the early days. Technical Note 13.1 has a brief discussion of how solar cells work and the types and efficiencies of cells in use.

Most of the installations so far have been of the standard silicon solar cell roof panels, but there is lots going on in the industry and new kinds of systems are being introduced. Figure 13.4 is an example of something different: a CIGS (cadmium–indium–gallium–diselenide) system developed by Solyndra Corporation. The elements are like long fluorescent light tubes with another tube inside coated with the PV material. The glass in the tube walls has a focusing effect and the system has a somewhat more uniform response during the daytime than flat panels.

PV installations are increasing rapidly, though from a small base. In the United States, California has the most aggressive program with a goal that its Governor calls "a million solar roofs." The aim is a capacity of 3 GW. However, just as with wind, there is a difference between capacity and output, this time driven by day-night and cloud cover. Figure 13.5 from the US DOE [34] shows the effective annual hours of output all over the United States (including both latitude and weather effects). There are 8760 hours in a year. If you live in Dallas, Texas, for example, you are between the 1800 and 1700

---

[5] California is running a pilot project in San Diego on solar hot water and will decide based on results what to do about incentives.

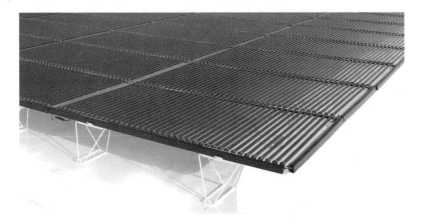

Fig. 13.4   Rooftop photovoltaic system. A Solyndra CIGS rooftop installation. (Courtesy of Solyndra, Inc.)

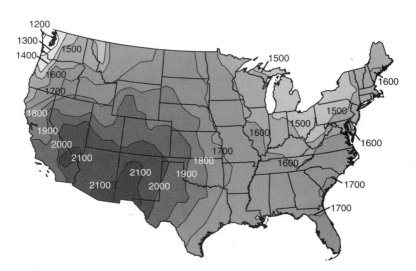

Fig. 13.5   Solar efficiency map. Effective hours of sunlight per year on a flat plate collector including weather effects. (*Source*: US Department of Energy's Energy Efficiency and Renewable Energy Division)

hour contours so your effective output is about 20% of peak and you will produce 1750 kilowatt-hours of electricity for every kilowatt of peak capacity installed. This map is for stationary flat plate collectors like those for California's million solar roofs. In California, Figure 13.5 shows that the average number of hours of collection is

1900 per year, so the energy delivered averaged over the year is only 22% of the capacity.

The time of peak demand in most places in the world is in the daytime when solar systems are most effective. PV systems can make an important contribution to reducing greenhouse gas emissions, but without energy storage there is no way to generate electricity from sunlight on a large scale during the day and store it for use at night. Small systems typical of residences can use simple battery systems for night-time storage, but PV solar is not yet suitable for baseload power (solar thermal electric systems can store energy – more on this later).

Typical systems for single residences have 3 to 5 kilowatt capacities and cost about $8000 per kilowatt. A 5 kilowatt system in California starts out at a cost of $40 000 (2008 costs). For this system the present federal tax credit is 30% or $12 000. California gives a rebate of $1.55 per watt or $7750. What started out costing $40 000 nets out after all the credits at $20 250. With all the rebates the payback time is typically about 15 years.[6] As time goes on there should be new types of photocells available at lower cost and PV systems may become more attractive.[7] As of now, they are only for rich countries.

Solar thermal electric generation is the second solar electric system. This technology is more typically used in systems much larger than those on roof tops. Figure 13.6 shows an example of a system installed in the state of Nevada. The curved mirrors form long troughs that focus the sunlight on a tube that carries a fluid that is heated to a high temperature. The long axis of the system runs north–south and the troughs rotate to follow the Sun from east to west.

In principle, if the insulation on the tube was perfect, the fluid could reach the same temperature as the surface of the Sun, 6000 °C. These systems are used to generate steam to run electricity-producing turbines and typically heat the fluid in the tube to 300 to 500 °C. The overall electrical efficiency is typically 25% to 35%, better than all but the most exotic (and expensive) solar PV cells. There are two other potential advantages. Tracking systems make more efficient use of sunlight. Figure 13.7 show the relative output of a flat plate and a tracking collector as a function of time of day. The graph is for the latitude of Los Angeles at the equinox. The tracking system snaps to

---

[6] The Sharp Corporation has a calculator at www.sharpusa.com/solar/home where you can see what happens as costs and interest rates change.

[7] California also has a program for large PV systems more typical of commercial applications. The rules are too complicated to summarize here.

Fig. 13.6   A solar thermoelectric system. Solar concentrator mirrors and tube system. (Photograph of ACCIONA's Nevada Solar One courtesy of ACCIONA)

full output almost as soon as the Sun rises while the stationary flat collector has a slow rise and fall, and only reaches full output at noon.

Solar thermal electric systems have the potential to store energy as well. The storage is as heat, not as electricity. A heat reservoir like the tank used in the solar hot water system, but with a fluid that can stand much higher temperatures, can be heated during the day and the heat extracted later to run the generating turbines. Since the time of peak electrical demand usually starts a few hours after sunrise and ends a few hours after sunset, this is an important advantage for the thermo-electric systems.

There are two disadvantages to thermo-electric systems. On a lightly overcast day, PV systems can still produce 30% to 40% of their normal output, whereas focusing systems produce nearly nothing. Also,

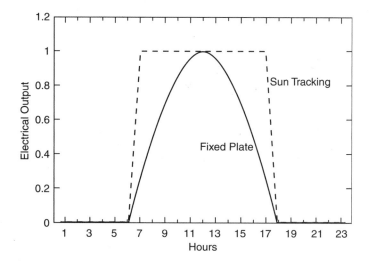

Fig. 13.7   Solar output as a fraction of peak as a function of time of day.

the added cost of the steam generation and turbine systems makes the thermoelectric system too expensive for smaller installations. Even so, in places where peak demand is in the summer as it is in most of the United States, solar electricity has a potentially large role to play.

13.4 GEOTHERMAL

The Earth has at its center an iron core compressed by the weight of 4000 miles of rock above it, and at a temperature somewhere between 4000 °C and 7000 °C (6400 °F to 11 200 °F). The high temperature was originally generated by the accumulation of all the material that make up our planet, and the temperature is maintained today by the heat from the decay of radioactive uranium, thorium, and potassium that were in the primordial material that came together 4.5 billion years ago. The heat from the interior of our planet leaks slowly to the surface, slowed by the insulating properties of the 4000 miles of material to be gotten through.

At the surface the heat flow from the interior is less than one five-thousandth of the heat from sunlight, and is negligible in determining the temperature of the surface. There are places where the very hot interior material comes close to or even on to the surface, and then we have hot springs or volcanoes emitting rivers of molten rock. Hydrothermal systems in these areas have been used for millennia for heat and in the past 100 years for electricity production. If you go far enough below the surface the temperature rises even where there is no volcanism, and if you go deep enough it becomes high enough to be used for energy production. The potential of this second geothermal source is only now beginning to be explored by what are called enhanced geothermal systems (EGS).

Strictly speaking, geothermal energy is not renewable. Almost all of the applications are mining the heat that is stored below the surface. The average heat flow to the surface is very small, only 0.057 watts per square meter, so to produce the 2000 MW of power that comes from the Geysers geothermal field in northern California would require collecting the natural heat flowing up from deep in the Earth from 14 000 square miles (35 000 square kilometers) of the surface. The collection area is much smaller than that so the local high temperature heat reservoir is being depleted slowly. Nevertheless, geothermal energy is included in the renewables because there is a lot of it and the total resource will last a very long time, though individual power plants will become exhausted.

Hydrothermal systems have been in use for as long as there has been life on Earth. The Rift Valley in Africa, thought to be the cradle of humanity, has been a refuge during the periodic ice ages that have punctuated the evolution of humanity. From the temples of ancient Greece, to the baths of Rome, to the spas of Europe, to the National Parks of the United States, hot pools of water have been important to mankind for relaxation, medicinal purposes, and the heat that is available (Iceland gets over 85% of the energy to heat its residences from its hot springs). The first use of geothermal energy for electricity production was in Italy in the early 1900s. Today, the single largest geothermal power plant is the Geysers in California.

The technology for electricity production is well developed and new plants are being built wherever there is a good source of hydrothermal energy. There are environmental issues, and some care has to be taken in developing the resource. The earliest plants simply mined the steam from underground and then let the cooled waste water out. It can be loaded with gases and minerals that are harmful. Also, the underground water itself is a limited resource and the early plants found that their output was decreasing as the underground water was depleted. It was economics more than environmental concerns that led to the development of closed cycles where the condensed steam, after use in the generators, is pumped back into the heat reservoir.

According to the REN 21 group [35], there are about 9.5 GW of hydrothermal electricity generated worldwide in over 20 countries. The United States has been the largest producer, but is being overtaken by the Philippines where there are large unexploited hydrothermal fields. Any place where there is significant volcanic action is a candidate; for example Iceland, New Zealand, or Japan. This resource will grow, but it represents only a minute fraction of TPES and will have to grow much larger to have a significant impact on emissions of greenhouse gases.

I was for a while involved with a proposal to produce large amounts of geothermal electricity from the volcanoes on the Island of Hawaii (known also as the Big Island). My laboratory in California and our sister lab in Japan were working on the design of a new kind of particle accelerator that needed a long tunnel and a lot of electric power. We looked at conditions there and found that on the Big Island tunneling conditions were very good for an underground accelerator some 20 miles long. What was lacking was power to run it. When we approached the governor's office we found a surprisingly warm reception. They had been looking at making geothermal power and transporting it to Maui and Oahu by undersea cable, but had no real use on the Big Island itself with its small population and lack of industry. We gave them a use which they hoped would make the residents enthusiastic about the project. Unfortunately it all came to naught when the cost of the undersea cable was found to have been greatly underestimated.

The new thing in geothermal is the program aimed at developing the technology to allow the mining of heat energy from hot dry rock at depths of five to ten thousand feet (3 to 6 km) in regions where drilling to that depth is not too difficult. The basic idea is shown in Figure 13.8. An injection well and an extraction well are drilled to the desired depth. Water is pumped down the injection well where it is heated to a high temperature, taken back to the surface through the extraction well, turned to steam to drive an electrical generator, and finally sent back down the injection well to be used over and over again. This approach was tried in the 1970s in a Department of Energy funded program, and failed; the fluid became contaminated with corrosive material, and getting the right conditions in the rock underground proved too difficult. It is being tried again in Australia, Europe, and the United States.

The key to making this work is the condition of the rock between the injection and extraction wells. It has to be appropriately

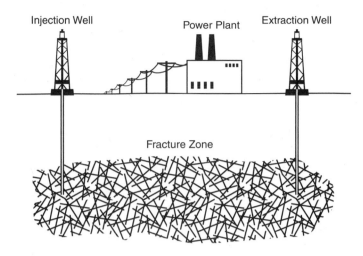

Fig. 13.8   An enhanced geothermal system.

fractured so that the water has many paths between the two that are spread out so that heat can be obtained from a large volume of rock. Remember that rock is a poor conductor of heat and if there were only a few paths the rock near those would quickly cool and little energy could be extracted in the long run. Even if the rock is well fractured there is the danger of what are called short circuits where in the midst of all the little cracks there are a few big ones that connect the two wells. In this case most of the water rushes through the short circuits, again limiting the amount of energy that can be economically extracted. There are two reports that are worth looking at if you are interested in more details. One is a 2006 study by MIT commissioned by the DOE's Idaho National Laboratory that looked at the potential of EGS energy [36]. I find this report to be an excellent primer on the technology, but too optimistic about the ease of overcoming the technical obstacles to large-scale deployment. The other is a 2008 analysis by the DOE itself of the MIT report that reviews the MIT assumptions [37].

Table 13.2 is from the MIT report and I include it to give an idea of the scale of EGS plants. The table gives the surface area required for a plant of a given electrical output and the very large deep subsurface volume of the fractured rock region required to produce the indicated electrical energy. They assume a 10% conversion efficiency of heat to electricity, and a re-drilling and creation of a new subsurface reservoir in a new area after five years of operation of a given well system to make up for the cooling of the rock.

Table 13.2 *Surface area requirements and subsurface fractured rock volume for EGS plans of a given electrical output*

Electric output (MWe)	Surface area (square kilometers)	Subsurface volume (cubic kilometers)
25	1.0	1.5
50	1.4	2.7
75	1.8	3.9
100	2.1	5.0

There are environmental concerns about EGS, particularly about microseisms or tiny earthquakes. The DOE had a plant at Rocky Flats near Denver that was part of its weapons complex. The plant began to dispose of its chemically contaminated water by injecting it deep underground. This seemed like a good idea at the time; careful environmental studies were done to see that there would be no contamination of the water supply of Denver or anywhere else in the region, and even the protestors were satisfied about that issue. What no one had thought of was the possibility of tiny earthquakes arising from the lubrication of the natural cracks in the rocks deep underground. After pumping had gone on for a while tiny tremors began to be felt in the region and these continued until, as an experiment, pumping was stopped. With the cessation of pumping came a cessation of micro-earthquakes. The advocates of EGS say this phenomenon is understood and they know how to control it. We will have to see if this is so.

Australia and Europe are the leaders in EGS technology with commercial-scale projects being constructed in both Australia and Germany. The United States, which started the development in the 1970s, is now behind and in need of a technology transfer injection. Too often in the United States the impatience of the Administration and Congress leads to the abandonment of science and technology areas that do not produce results at first try. In spite of the uncertainties I do agree with the main MIT conclusion: the promise of EGS is enough to make a large-scale try worthwhile to evaluate the technology and the costs.

One other approach is worth mentioning: heat pumps. These are small systems that are capable of supplying both residential heating in winter and cooling in summer. If you drill down only a few tens of feet in almost all regions of the non-tropical regions of the world, the underground temperature will be about 50 °F (11 °C). If you wonder how a reservoir at only 50 °F can allow you to heat your house, you should look at your refrigerator. In the refrigerator heat is extracted from the inside of the box by a refrigerant and dissipated into the room by the cooling coils and fins on the back of the standard refrigerator. In the heat pump, heat is extracted from the ground, cooling the ground slightly, and dissipated in the house by a system very much like what is on the back of a refrigerator. It can also be run in reverse to supply cooling. These units are reliable and cost-effective. Their contribution to reducing emissions is small today but is growing rapidly.

13.5 HYDROPOWER

The contribution of hydroelectric power systems to world renewable energy was about 340 gigawatts averaged over the year (GWa) in 2006,[8] dwarfing all but biomass. In the United States, hydropower delivered 33 GWa which amounts to over 80% of the electricity from all renewables. Worldwide, almost all of this comes from large power dams like the Columbia River dams in the United States, the Aswan dam on the Nile River in Egypt, or the Three Gorges Dam on the Yangtze River in China. Large power dams have potential environmental problems of their own, and a balance will have to be struck between those concerns, concerns about global warming, and the need for affordable energy. In the developing world the need for energy is dominant.

Big power dams store water as well as generating electricity and more of that will be needed as temperatures rise. For example, California gets much of its water in the summer from melting snow in its mountains. As temperatures rise, there will be less snow and more rain with more runoff in the winter, meaning less water in the summer unless storage is available. This same problem exists in many countries, and each will have to make its own choice. It is unlikely that another big dam will be built any time soon in the United States or Europe, but Asia has made different decisions based on their need for energy and perhaps the world's need to limit emissions from fossil fuels. At the moment, according to the IEA World Energy Outlook 2008, hydropower capacity under construction amounts to 167 GW

---

[8] River flow varies over the year so hydropower is generally given as a yearly average. The capacities of the plants are typically about twice the average output.

of which 77% is in Asia. China leads with 93 GW followed by India with 15 GW.

Smaller hydropower facilities generate less environmental concern and there seems to be a significant potential, at least in the United States, where a detailed study has been done by the DOE. Hydropower is broken down into four classes: *large* are those that can produce more than 30 MWa, *small* are in the 1 to 30 MWa class, *low* are less than 1 MWa, and *micro* are less than 0.1 MWa. A recent study by the DOE evaluated all low and small hydropower sites in the United States [38]. They surveyed over 500 000 potential sites and after applying their selection criteria came down to 5400 with an electrical potential of 18 GWa. Whether any of these will actually be developed depends on costs.

Hydropower systems can be easily turned on or off so that electricity can be generated when needed. That capability leads to an application of hydropower called pumped storage. The idea is that with a lake at some high elevation and another at a low elevation, electric power can be generated by letting the water run downhill when needed, and then pumping the water back uphill to be used again when the electricity is not needed. With the right kind of equipment, the same turbines that generate electricity from falling water can be run as pumps to send the water back to the top. Turbine efficiencies are about 90% for generation and can be as high as 80% when run as pumps. Of course, the pump power has to be supplied from other sources.

You do not need a large river to generate large amounts of power for part of the time. All you need is a big enough lake and a water supply that can make up for losses. This makes sense where daytime power costs more than night-time power so that the supplier can make a profit by playing off the difference; generating electricity in the daytime and pumping at night. As far as greenhouse gas emissions are concerned, they depend on those of the emissions of the power source used to pump the water uphill. Solar power does not make much economic sense for this since it is generated during the day when electricity prices are high. It does not make sense to use high-priced daytime power to pump water uphill to generate power at night when prices are low.

There is little more to say about hydropower. The technology is mature and its use will be determined by choices made in many different places. Costs for large-scale hydropower are at grid parity or below, and as far as greenhouse gases are concerned, it is a winner.

13.6 OCEAN ENERGY

In Chapter 8 the total energy in all the world's waves and all the world's tides was estimated to be about four times the TPES. Of course only a tiny fraction of that is really accessible. The tidal lift or waves or the temperature difference between surface and deep waters in mid-ocean are no use to us because we cannot yet manage to get the energy harvested back to the shore. Nevertheless, there is a source of potential significance near the shore and much work is under way trying to harvest part of it. There has been little success so far and only about 10 megawatts are being generated compared with a total world electricity generation of over 2 terawatts. A colleague at Stanford who works in the field tells me that only about $10 million per year in revenue from ocean-generated electricity sales is being produced after a total world investment in the technology of about $500 million over many years. However, hope springs eternal in the human breast and that plus generous subsidies have revived interest in trying to do better than was done in previous attempts to harness waves, tides, and ocean currents.

There are far too many different types of devices being proposed or under test for me to attempt to describe them here. There are books that go into some detail on wave and tidal energy as well as on other forms of renewable energy.[9]

The oceans, even near the shore, are a hostile environment. Any system has to cope with the corrosive properties of salt water, and the wildly varying conditions ranging from calms to violent storms. Unfortunately for system costs, the design has to be able to survive the worst of storms, and that alone make all ocean systems expensive. Still, there are isolated communities and islands where an ocean source would be economically worthwhile. Wave-, tidal-, and current-driven electrical generation systems will be worth another look in five to ten years to see if any of them have overcome the difficulties that have led to defeat in the past.

13.7 THE ELECTRIC POWER DISTRIBUTION GRID

Electric power is brought from the power generating plant to the end user through a complex system of high-voltage, medium-voltage,

---

[9] One I like is Godfrey Boyle's *Renewable Energy* [40]. The book was designed for an undergraduate course on renewable energy and has more mathematics than the general reader might like, but it also is descriptive and worth more than a casual look if you are interested in learning more.

and low-voltage power lines that are collectively known as the grid. The grid was never "designed" in the sense that a group of sophisticated engineers looked over the entire county's collection of power plants and load centers and laid out an optimized system of wires to connect them all. On the contrary, it mainly just grew from what we had years ago by adding a patchwork of transmission lines to get the power to load centers that changed over time. Generating plants used to be near cities and the grid was mainly local. Power plants were then moved away from cities as real estate values went up, and the grid began to stretch out. Regional power centers came like the giant hydropower systems of Niagara in New York and the Columbia River of the Pacific Northwest, and the grid became regional. It was never designed to move electricity for long distance, but this is what it is increasingly called on to do. It can only do it through an increasingly fragile patchwork of interconnects where the failure of one line can bring down the electric supply of an entire region of the country.

In the United States the largest wind resource seems to be in the Great Plains, while the best solar is in the southwest. The same sort of situation exists in many places. The UK as mentioned will have its largest wind farms in the north while the load is in the south. If these wind resources are to have more than a regional impact on decarbonizing the electricity system, the power has to be moved a long way and the present grid cannot handle the projected loads. Therefore, there is a lot of talk about a new grid, but very little discussion of what it is to do. Here are a few options:

- Move large amounts of power from the wind farms of the Great Plains and the solar installations of the southwest to the country's major load centers;
- Make the system more robust so that the failure of one major line does not seriously affect a large area;
- Have the system capable of using power from distributed small systems like the million solar roofs program in California.

I could continue the list with smart metering, very fast load adjustments, etc., but the point is that the functionality has to be specified in advance or at least the desired upgrade path laid out if only the first objective is to be met. No organization has been assigned the responsibility as yet. Construction will be a problem too. Currently, it takes about 10 years to get approval for a major power line route.

A model that might be used is that of the US interstate highway system begun in the 1950s during the Eisenhower administration. It had been designed long in advance mainly by the military as a way to move goods and troops in the event of another major war. Since it was cleverly dubbed the National Defense Highway System, the federal government laid out the route in consultation with the states and built it with federal money. Maybe if everyone called their new grids a National Defense Electron Highway System we could actually get them built in a reasonable amount of time.

Technical Note 13.1:   Photovoltaic cells

In the semiconducting material from which PV cells are made, electrons are stuck onto atoms and cannot move freely in the material; they are stuck in what is called the valence band of the material. To get them to move they have to be shifted up in energy to the conduction band where they can move through the material and out onto a wire to deliver electricity and then return to the semiconductor. Where they leave and where they return to the PV cell is determined by specific impurities that are introduced in the manufacturing process.

There is a minimum amount of energy called the band gap that has to be delivered to move an electron up into the conduction band. If the quantum of light (photon) has less energy than the band gap, the electrons stay stuck. If it has much more, the electron is excited above the band gap and the excess energy is lost to heat rather than being converted to electricity. The band gap in silicon is about 1.1 electron volts. The band gap for a single material system that gives the highest conversion efficiency is about 1.5 electron volts when the solar spectrum is considered. With silicon only, the maximum possible efficiency is about 30%, but nothing gets near that.

Silicon solar cells made from single crystal material like that used for computer chips have typical efficiencies of about 15% and are the most expensive. Polycrystalline materials have efficiencies of about 10% and cost less to make. Amorphous thin film cells have efficiencies of about 8% and are the least costly of all the silicon-based cells. In the overall economics the thin film cells are the front runners. Other materials include gallium arsenide, copper indium selenide, copper indium gallium diselenide, and

Technical Note 13.1 (*cont.*)

cadmium telluride. They have different band gaps, and different efficiencies.

The highest efficiency solar cells are laminates of different material whose band gaps match different parts of the solar spectrum. The highest efficiency reported for one of these is 40% but they are very expensive now. Their proposed use is with concentrator systems so that focusing mirrors can get the sunlight from a large area onto a small cell. Some say that nanotechnology has the potential to lower the cost of multi-band-gap systems, but no such devices have yet been made.

14

Biofuels: is there anything there?

My first introduction to the idea of biofuels came when I met the Nobel Laureate chemist, Melvin Calvin, in the late 1970s (his prize was awarded in 1961 for the discovery of how photosynthesis worked). It was the time of the Arab oil embargo and he had a dream of what he called growing oil. He had found a plant in the Amazon that produced oil that could directly substitute for diesel fuel, and was working on improving the output of a different plant that could grow in the temperate zone, and on poor ground. He wanted, through genetic engineering, to greatly increase its natural production of an oil-like substance. He did not think using food crops for energy systems was a good idea because of population growth. We would need all the food we could get. Mel retired in 1980 (continuing to work as do most of us) and died before he succeeded. The science community is still trying to bring Mel Calvin's vision to life.

Today, in the United States biofuels means ethanol from corn, while in Brazil it is ethanol from sugarcane (the European Union has an ethanol program too, and I will come back to it). After looking in some detail at the US program, I confess that I have become a biofuels skeptic. Most of what one hears about corn as a source of fuel ethanol that saves energy and reduces greenhouse gas emissions is propaganda from agribusiness (I think Calvin would agree). Sugarcane is one crop that does give the promised benefits, but even its long-term contribution to reducing greenhouse gas emissions depends on how land is used. There is an intensive worldwide research program aimed at developing much more effective biological sources of fuel, but it has not yet reached practicality.

Here, I will discuss what is happening now and what might happen in the future. Even if the new methods do prove to be effective, biofuels cannot be the entire answer, nor should they have to be. As I said earlier, there is no one magic bullet that can solve all of our problems.

The theory behind the hope for bioenergy is simple. Plants get the carbon used for growth from CO_2 taken out of the air, turning the carbon into stems, leaves, and fruit while returning the oxygen to the atmosphere. Sunlight provides the needed energy to drive the photosynthesis process that does the job. In a sense all bioenergy is a form of solar power. If plant material can be efficiently made into fuel, burning that fuel only returns to the atmosphere what the plant removed in the first place and there is no net emission of greenhouse gases, but that is not the case with today's biofuels.

Roughly 1.6 billion poor people on our planet, who have no access to commercial energy, gather and burn plant material for cooking and heat. They live according to the theoretical model and generate no excess greenhouse gas from their biofuels. As long as they use only what grows naturally the model is correct. However, modern farming requires extra chemical and energy inputs, and here is where care is needed to identify all the inputs and include them in balancing the energy and environment scales. Energy is required to produce the needed fertilizer, run farm machinery, harvest the crop, transport it to the factory, and run that factory. Greenhouse gas emission and energy consumption have to include all of the inputs to the process.

There also are unintended consequences as food crops are turned into fuel crops. These include increases in food prices and competition for land and water. With world population forecast to increase from six billion in the year 2000 to about nine billion in 2050, the food-into-fuel program will have ever more serious unintended consequences, particularly on prices, and become unsustainable as now practiced.

Ethyl alcohol (ethanol) derived from plants can be used as a motor fuel as well as for convivial drinking. Alcohol biofuel has been trumpeted as a way to greatly reduce emissions of greenhouse gases in the transportations sector, as well as a way to reduce demand for the oil that now is mainly used for transportation. In today's biofuels program, starches in corn and wheat are first converted to sugar, or sugar comes directly from a plant like sugarcane. The material then goes through fermentation to produce alcohol. This is basically the same process that has been used for thousands of years to make the alcohol in wine, beer, or whiskey [41].

I will look first at today's major biofuels programs which I will call Phase-1 programs, focusing on those of the United States and

Brazil. The US program now is based on corn as the source material; the Brazilian program is based on sugarcane. The US program is of marginal utility while the Brazilian program is highly effective.

A more effective system for producing biofuels is under study. This Phase-2 program is aimed at making practical a system called cellulosic ethanol where the entire plant is used, not just the starches and sugars. There will still be problems with competition with food crops and land and water use, though these are claimed to be less severe. Phase-2 is still a research and development program; there is no commercially viable process yet. I will also briefly mention the possibility of a Phase-3 where the inputs are different and the outputs can be fuels that are different from alcohol.

14.2 PHASE-1: ETHANOL FROM STARCH AND SUGAR

The effectiveness of biofuels relative to gasoline (or diesel fuel) can be measured in three ways. One way compares greenhouse gas emissions taking all the emissions from the energy used in biofuel production into account. On this basis only sugarcane-based biofuels reduce greenhouse gases by a large amount compared with gasoline or diesel fuel on an equivalent energy basis.

A second way is to compare the total energy used to make ethanol to the energy content of the ethanol itself. Here, too, only sugarcane gives a significant benefit.

A third way looks at costs relative to gasoline. A gallon of ethanol contains less energy than a gallon of gasoline, and costs should be compared on an equal energy basis. It takes 1.4 gallons of ethanol to equal the energy in a gallon of gas so be careful about comparing costs. Prices are volatile as the oil prices and ethanol feed stocks move up and down. On this ground, ethanol from corn is sometimes a winner and sometimes a loser while sugarcane ethanol is a winner most of the time.

Current US law requires that ethanol be blended in with gasoline. The ethanol mandate dates originally to the Farm Bill of 2002, superseded by the Energy Policy Act of 2005, in turn superseded by the Energy Independence and Security Act of 2007. The newest law sets a goal of 9 billion gallons of ethanol to be used in the year 2009 and 36 billion gallons in 2022. It defines conventional biofuels as those derived from cornstarch. New ethanol factories have to have a reduction in greenhouse gas emissions of at least 20% compared with conventional motor fuel. The Environment Protection Agency (EPA) can reduce this requirement to as low as 10% if it determines that meeting the 20% goal is not feasible. There is already an argument over what to include in counting the greenhouse gas emission from ethanol. Including the effect of land-use changes makes it very difficult for ethanol to meet the 20% reduction standard. I would guess that the goal will be reduced for most factories.

Advanced biofuels come from sources other than cornstarch, and they have to meet a goal of no less than a 50% reduction of greenhouse gas emissions (EPA can reduce the requirement to 40%). Cellulosic ethanol will be discussed later but it has to have a 60% greenhouse gas reduction (50% is its minimum). Cellulosic ethanol is not now available. There are demonstration facilities struggling to come up with a commercially viable production system.

Ethanol production in the United States gets a tax credit of 51 cents per gallon and foreign ethanol has an import duty of 54 cents per gallon levied on it, mostly to keep Brazilian ethanol out of the United States. The Brazilian program is discussed in more detail below, but it is much more effective in greenhouse gas reduction and in reducing energy use than the corn-based material required in the United States today.

When all the energy inputs in making corn ethanol are included, starting at the fertilizer factory and ending at the ethanol factory's loading dock, greenhouse gas emissions range from a few percent worse than those of gasoline to about 15% better on an equal-energy content basis (remember it takes 1.4 gallons of ethanol to get you as far as 1 gallon of gasoline will). Energy inputs in making ethanol are typically about 90% of the energy content of the ethanol produced. The greenhouse gas reduction depends on the energy sources used in the fertilizer-through-factory chain. The numbers above are based on the average mix of fossil, renewable, and nuclear energy used in the United States today. In the Pacific Northwest, for example, the main source of electricity is hydroelectric, and the greenhouse gas advantage would be greater. For regions where the main source is coal it would be worse. There are many calculations of the benefits of ethanol-based biofuels. I have looked at many and find that assumptions and methodology vary. The best that I have found is an article in the journal *Science* [42], and the numbers above are based on it.

The typical bushel of corn produces 2.8 gallons of ethanol. The total corn crop in the United States amounts to around 11 billion bushels per year, though this is variable depending on the weather. The 9 billion gallon ethanol mandate for the year 2009 requires the use of about 30% of the corn crop, and the 36 billion mandated for the year 2022 would require more than the entire corn crop (115% of it). Either advanced biofuels will be required to meet the future mandates or much more land will have to be converted to growing corn.

A recent report from the National Academy of Sciences on water use [43] says, "If projected future increases in use of corn for ethanol production do occur the increase in harm to water quality could be considerable. Expansion of corn on marginal lands or soils that do not hold nutrients can increase loads of both nutrients and sediments. To avoid deleterious effects future expansions of biofuels may need to look to perennial crops like switch grass, poplars/willows or prairie polyculture which will hold the soil and nutrients in place."

As mentioned earlier the cost of food is going up, and advocates of corn ethanol say that it is not their fault. They are partly correct. The price of corn doubled before the US Midwestern floods of the spring of 2008. That price increase is mainly their fault. After the floods the price of corn increased by another 50%, which is surely not their fault. The Governor of Texas in the summer of 2008 asked the EPA to reduce that state's requirement for corn ethanol use so that more corn will be available for animal feed. The hog farmers of Iowa are at odds with the corn growers of Iowa because feed prices are so high. There is a final bit of irony: the increase in corn prices has led to the abandonment of some plans to build more ethanol factories.

You may wonder with all of this if the US policy on ethanol makes any sense. Technically it makes no sense at all unless you believe that reducing oil imports is the only way to measure the benefits. Since the energy inputs required to make ethanol are about the same as its energy content, using ethanol as a motor fuel is really driving your car on a combination of coal, natural gas, nuclear power, and oil (don't forget farm machinery and trucks to transport the crop) depending on the mix of energy sources where the corn is grown. There are cheaper ways to convert coal to liquid fuel though they tend to produce more greenhouse gas than gasoline. Converting natural gas to liquid fuel or using it directly is better, but we import a lot of that too. Also, it may seem odd to you, as it certainly does to me, to mandate an increase in ethanol production by the year 2022 that can only be met with a process which has not yet been developed.

Where the US ethanol program does make sense is politically. Its origins were in the Farm Bill of 2002. Enthusiasm in Washington for corn ethanol increased dramatically in the run-up to the US election of 2004. The goal was to capture the votes of the corn-belt states, and both political parties jumped in. The result was that no one got any political advantage and the politicians are now afraid to back away from what, in fact, is a not very useful but very expensive program.

I have dealt with members of Congress and the administration for many years and there are very smart people of both parties in both places. How we got where we are on these issues, as well as in others involving complex technical questions, can in part (I think a very big part) be traced back to a fateful step taken by Congress itself in 1995, one that I have mentioned before. When the Republicans took control of Congress in 1994, one of the first actions in what is called the Gingrich revolution was to abolish the congressional Office of Technology Assessment (OTA). It was created in 1972 to provide advice to congressional committees on complex technical issues. While OTA had become pretty slow in response by the early 1990s, Congress would have been much better off fixing it rather than killing it. Congress now has no source of independent technical advice to add to what it gets from lobbyists and from reports it requests from the National Academy of Sciences. Lobbyists have their point of view and it is important to listen. The NAS tends to be slow because of the rules on vetting the appointment of members of study groups, but it will produce a good report in 1.5 to 2 years. Timely technical advice on important problems is needed. The techies should not have the last word; political issues are an essential part of decision making in a democracy, but technical input is needed so that the politicians understand the consequences of what they do. OTA is technically not dead; it just has no money for people, phones, computers, stationery, etc. I hope it, or something like it, is brought back to life.

The Brazilian program is different. Brazil makes its ethanol from sugarcane, a crop much more efficient than corn at turning sunlight and CO_2 into material that can be fermented into alcohol. Roughly a third of the sugarcane plant is leaves and stems that are left behind in the field after harvesting, a third is a liquid very high in sugar that is pressed out of the remains of the plant, and a third is the post-pressing leftovers called bagasse. The bagasse is burned to generate power to run the ethanol factory, which gives a big advantage to sugarcane over corn in energy used to make ethanol since the energy comes from the sugarcane plant itself and not from something external. In Brazil, bagasse typically generates more power than needed to run the ethanol factories and the excess is sold back to Brazil's electrical utilities.

The Brazilian program is more than 30 years old. It started in the 1970s when the first OPEC oil embargo and the Iranian Revolution worried the Brazilian government about national security and Brazil's dependence on imported oil. The government subsidized the development of the ethanol industry directly and through a requirement that all auto fuel had to be at least 20% ethanol. At the 20% level, ethanol mixed into fuel requires almost no adjustment of the auto's engine. The ethanol industry in Brazil is now mature enough to run without any government subsidies, unlike the case in the United States with corn. All Brazilian sugarcane mills produce both sugar and ethanol and adjust the balance between the two as world prices for oil and sugar change.

In 2007 Brazil produced about 5 billion gallons of ethanol, about 20% of which was exported, including some exports to the United States. Brazilian car manufacturers make "flex-fuel" cars that can run on any mix of ethanol and gasoline; sensors determine the blend of fuel in the tank and feed this information to the engine-control computer which makes the adjustment required to use whatever blend it is. Most of the cars produced in Brazil now are these flex-fuel vehicles. Ethanol makes up about 40% of automobile fuel and 20% of total fuel used for all vehicles including heavy trucks.

The ratio of energy contained in sugarcane-based ethanol to the energy required to produce it is roughly 8-to-1, far more favorable than US corn ethanol. Ethanol yield per acre is about 800 gallons, approximately 2.5 times the yield per acre of US corn ethanol. Greenhouse gas emissions are only about 20% of those of gasoline.

Brazil uses 1% of its arable land for cane growing, almost all of it far from the rain forest. This is an important point for greenhouse gas emissions. Forests sequester carbon dioxide and store a lot of it underground as well as in the trees themselves. Conversion of forest land to use for any annual crop releases large amounts of greenhouse gases which continue over a long time. Changes in land use already account for 15% of all world greenhouse gas emissions, but this does not seem to be a problem yet for ethanol in Brazil.

The bottom line is that Brazilian ethanol is sustainable and not subsidized at the current level of blends with gasoline. US corn ethanol is not sustainable and heavily subsidized (51 cents per gallon for a subsidy, plus the 54 cents per gallon tariff to keep out imported ethanol). The mandated amount of alcohol required for blending for gasoline by US law in the year 2022 is not doable without some new source which might even come from Brazilian sugarcane.

Despite the success of the Brazilian sugarcane program, Phase-1 biofuels do not offer a solution to global problems. A recent analysis of the world potential for all Phase-1 biofuels gives the whole, depressing story [44]. Taking all of the worldwide production of corn, sugarcane, palm oil, and soybeans, and converting it into liquid fuels gives a net energy equal to about 1% of today's global primary fossil-fuel energy. It can have no significant worldwide impact on greenhouse gas emission, but can make sense in special cases like Brazil. A large expansion of Phase-1 biofuels production can have only bad effects on food prices with little if any effect on climate change.

14.3 PHASE-2: CELLULOSIC ETHANOL

Plants contain much more carbon than is recovered or converted even from sugarcane. The walls of the cells that make up the structure of a plant are composed of materials called lignin, cellulose, and hemicelluloses locked in a complex tangle of fibers. The individual fibers are less than one-thousandth of the diameter of a human hair, and each is composed of a bundle of even smaller long-chain molecules. The aim of the cellulosic ethanol program is to find an efficient and affordable process to get at this material and turn at least the cellulose and hemicelluloses into ethanol. Breaking down the lignin would be a bonus. Termites do it, so why not us?[1]

Success would mean a huge increase in material that could be turned into ethanol or even used directly for energy production like bagasse in Brazil. Development of cellulosic technology is being supported by private industry as well as by the government. Pilot plants are being built to test process and costs at commercial scale. A joint study by the US Department of Energy and Department of Agriculture calculated that more than a billion tons of dry plant material per year would be available in the United States [45].

There are, of course, cautions; we do not want to get into a regime as bad as corn ethanol with all of its deleterious effects on food prices and land use. The hope for the cellulosic program is that crops can be grown on land that is only marginally useful for agriculture. The focus is on fuel crops that are perennials like grasses and forest waste rather than annuals like corn. Perennials sequester CO_2 underground in their ever growing root system. Converting a field from a perennial to an

---

[1] The International Energy Agency has a good status report on second-generation biofuels. It is available at http://www.iea.org/textbase/papers/2008/2nd_Biofuel_Gen.pdf . They estimate that it will be 2015 before a commercial demonstration is done.

annual crop increases emissions as the carbon trapped underground is slowly released. It takes many years to get it all out and in that period even cellulosic ethanol will have larger greenhouse gas emissions than gasoline. A recent study of the effect of land-use change illustrates the problem [46]. For example, converting land to corn for ethanol production increases greenhouse gas emissions by 50% compared with gasoline. It takes a long time for the small greenhouse gas gain from ethanol to balance the loss from land-use change.

There is an increasing amount of research aimed at understanding the effects of biofuels on land use, water, and food prices. This is to the good if anyone pays attention. But there still is no comprehensive integrated analysis that specifically looks at the negatives as well as the positives, and I therefore remain a skeptic on cellulosic ethanol until this kind of work is done. We already know that Phase-1 biofuels are losers except for sugarcane, and even sugarcane will become a loser if the growers convert the wrong kinds of land to its use. As to food prices, farmers are business people and they will use the best land to grow the crops that bring the most income. If that crop is a fuel crop that is what will be grown. Food prices will adjust to match those of energy crops.

14.4 PHASE-3: OTHER PROCESSES

There are a multitude of possible processes other than conversion to ethanol that can use carbon-based materials for fuels. Some are old and some are new.

The Fischer–Tropsch process is one of the oldest to be used on any large scale. It was developed in Germany in the 1920s to convert coal to a liquid fuel. It begins with the production of what is called syngas, a mixture of hydrogen and carbon monoxide, from coal and high-temperature steam. Further chemical reactions can yield liquid fuels. The use of liquid fuel produced from coal emits more greenhouse gases than gasoline. However, coal is not a necessary input though carbon is, and the carbon can come from plant material. The coal-based process is in use today and some are interested in making an efficient version based on plant material.

Pyrolysis is a process that heats carbon-based materials in the absence of oxygen and can produce in its first stage the same kind of syngas that comes from the Fischer–Tropsch process. If plant material supplies the carbon, it does not enhance greenhouse gas emissions. What comes after that depends on the rest of the process. Today, it is mostly the syngas that is used, but, as in the case of syngas from coal, the process can go on to produce liquid fuels. For those interested in what is happening worldwide, enter Pyrolysis in the Google search box and you will find many companies working in the area.

Our familiar natural gas, methane, is produced from waste in landfills and manure ponds. In some places this gas is being captured and either fed into pipelines or used directly for local energy production. This is a winner in the greenhouse gas fight because methane is an even stronger greenhouse gas than CO_2. The conversion of methane into CO_2 reduces the greenhouse effect in two ways. First, it reduces the impact of the gas emitted. Second, it displaces other fossil fuels that would have been used to produce the energy made from the methane. Localities and large farms are increasingly turning to using the methane from waste rather than letting it go into the atmosphere.

Bacteria and algae have a potential that is only beginning to be explored. There are biological systems that directly convert sunlight and nutrients into useful energy products. One example is the conversion of water into hydrogen and oxygen driven by sunlight and a modified photosynthesis cycle. The problem with this and many similar cycles is their inefficiency, but the hope is that bioengineering will improve them.

14.5 SUMMARY

In discussing the Phase-1 programs I have focused on corn and sugarcane. There are other sources as well and the conclusions about them are not significantly different. The European Union is the world's third largest producer of biofuels and by far the largest producer of biodiesel fuel. The EU Directorate for Trade and Agriculture has recently produced a report evaluating its biofuels programs.[2] Its ethanol production is based on wheat and sugar beets as input, and the conclusions are not different from what I have said about the corn and sugarcane programs. Wheat-based ethanol is more expensive than gasoline and its energy and greenhouse gas emission benefits are somewhat larger than corn, but much less than sugarcane or sugar beets. Sugar-beet ethanol is economically much better than wheat and its energy and greenhouse gas benefits are positive, though not as positive as those of sugarcane.

[2] Economic Assessment of Biofuels Support Policies, www.oecd.org.dataoecd /19/62/41007840.pdf

The EU analysis examined the effect of current support policies on greenhouse gas emissions in Canada, the EU, and the United States. They conclude that eliminating all price supports, tax credits, mandates for use, and tariffs would increase emissions by 0.5% to 0.8% above those expected for transport in 2015 with all the subsidies and credits. This comes at a cost of $25 billion for the period from 2013 to 2017 or $960 to $1700 per metric ton of CO_2e saved.

A fraction of the money spent on subsidizing a bad program might be used to develop some good programs including, perhaps, effective Phase-2 and Phase-3 programs.

15

An energy summary

Part II has covered the energy scene from fossil fuels through all the renewables. Part III is on policy, both for the United States and internationally. However, policy has to be based on reality, so I want to summarize the important technical points. I have tried in Part II to present all the facts without prejudice, but here I will let my own opinions shine through and end this chapter with a repeat of the energy scorecard shown earlier.

We are in a race to reduce global emissions of greenhouse gases while energy demand is going up fast, driven by two things: a projected 50% increase in population, and an increase in world per capita income. Continuing on our present course, world primary energy demand is expected to double by 2050 and double again by 2100. There is enough fossil fuel to let the world do that at least through 2050, but beyond then the supply of fossil fuel is not so clear. It is not just climate change that should be moving us away from our dependence on fossil fuels.

There are three roads to reduced emission: doing the same with less (efficiency), capturing (putting the emission away somewhere else than the atmosphere), and substitution (replacing fossil with non-emitting or low-emitting fuels). We need to use all of them and remember that the goal is emissions reduction, not merely replacing fossil fuels with the limited collection of things that are called Renewables. We have to keep in mind that advanced technology development is not just for the rich countries, but must be affordable and usable by the poorer ones too. It is not possible to reduce greenhouse gas emissions far enough without action in the developing world, even with zero emissions from the industrialized nations.

Some context relating goals to emissions is needed. I am writing this chapter in early 2009 when the new US administration is becoming engaged in emissions reduction strategy, and several goals are being discussed: returning emissions to the 1990 level by 2020 or 2030, and possibly going on to achieve a reduction to 80% below 1990 emissions by 2050. All the industrialized nations continue to get a bit more efficient each year. Energy intensity (energy needed to produce a dollar of GDP) in the US economy has been declining by about 1% per year for some time and that is projected to continue into the future. However, the economy has been growing faster; US primary energy use has been rising by about 1.2% per year and much faster than that in the developing world. In a business-as-usual scenario, US emissions will rise with fossil fuel use, from an actual 83 Quads of primary energy in 1990 and 101 Quads in 2006, and on to a projected 119 Quads in 2020, 135 Quads in 2030, and 170 Quads in 2050. According to EIA data, in 2007 only 10 Quads of emission-free primary energy were produced in the United States, and they were dominated by biomass, big hydro, and nuclear electricity generation.

The renewables, which are receiving so much attention, make up a microscopic fraction of world energy. There are two questions to be answered about them: can they be scaled up, and if scaled up are they affordable by anyone but the rich countries? My feeling is that the answer to both questions for wind is yes, though I am still concerned about the analysis of the required back up to maintain some sort of steady output. For solar photovoltaic my answer would be no on the basis of cost, unless some new kind of PV material comes from the world's laboratories. For solar thermoelectric and enhanced geothermal my answer would be maybe – it is too soon to tell. For ocean systems, I would say they are unlikely to contribute much. We would do much better at greenhouse gas reduction if that was the focus of the effort rather than only on expanding the use of renewables, as seems to be the case now.

Figure 15.1 shows what has to be done in the United States to meet the 2020, 2030, and 2050 goals. The height of each bar is the total primary energy supply (TPES) under a business-as-usual projection. The shaded parts show actual emission-free primary energy in 1990 and 2006, and what is required from them to match 1990 emissions in 2020 (43 Quads) and in 2030 (59 Quads), and to be at 20% of 1990 emissions in 2050 (155 Quads).[1] Part of the emissions reduction can come from efficiency measures that reduce demand; the rest has

---

[1] What has to be done is to displace sources with emissions with sources without such emissions. Most of the non-emitting technologies are designed to produce electricity. A very rough guide is that for every Quad of fossil-generated electricity that is replaced by solar, wind, geothermal, nuclear or hydro, primary fossil energy is reduced by 3 Quads because of the way the bookkeeping is done.

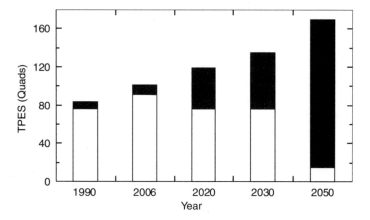

Fig. 15.1 TPES for the US projected to 2050 under a business-as-usual scenario. Shaded areas are reductions required to meet various goals. The TPES numbers for 1990 and 2006 are DOE EIA actuals. The shaded areas for 1990 and 2006 are actual emission-free energy including nuclear, big hydro, and renewables. The totals for 2020, 2030, and 2050 are EIA projections. The shaded areas show the emission-free contribution required to match 1990 in 2020 or 2030, or to be at 20% of 1990 in 2050.

to come from emission-free or low-emission sources. Making it even harder is the difference between output and capacity in solar and wind power. All the wind, solar, and geothermal power generated in 2007 totaled 0.2 Quads. Even the simplest goal, a replacement of 43 Quads of energy with the renewables by 2020, is a huge problem. All the tools in the chest will have to be used, including nuclear power, and carbon capture and storage if it can be shown to work. We can do these things, but how long it will take and how much it will cost is uncertain. It will not be easy.

Most of what is discussed about emission reductions in the press and by governments is focused on fossil fuel use which contributes 70% to the emissions that drive global warming. The other 30% comes from agriculture and land-use changes, and in the rush to do something about the 70% it seems to have slipped almost everyone's mind that the agriculture sector needs to be addressed too, something that gets more difficult as population increases from six billion in 2000 to the expected nine billion in 2050. I only touched on agriculture and land use in the chapter on biofuels, but these need attention that they are not getting. Some of the most aggressive goals for emission reduction that have been discussed (20% of 1990 emissions by 2050, for example) are impossible to achieve without doing something about agriculture and land use. It does seem absurd to worry about controlling emissions from airplanes which account for a few percent of global emissions as the EU is doing now, while leaving the 30% from agricultural and land use unregulated.

Efficiency in electricity generation was discussed in Chapter 10. US coal- and gas-fired generating plants are a long way from world leaders in efficiency. If all were world class, total US emissions would be reduced by about 5%. If all electricity generation was emission-free, emissions would go down by 38% (EPA numbers for 2006). If an old 1 gigawatt coal-fired power plant is replaced by a modern gas-fired power plant, 6 million tonnes of emissions would be avoided. However, there are no economic incentives now for merely eliminating 70% of the emissions from a coal plant. Cap and Trade or emission fees (discussed later) are supposed to move the system in this direction and will be winners if properly designed.

End-use efficiency was the subject of Chapter 11 where both transportation and buildings were analyzed. In the light-vehicle sector (cars, SUVs, minivans, and pickup trucks), the new US CAFE standard of 35 mpg by 2020 is a winner. It would be a bigger winner at 50 mpg in normal or hybrid vehicles by 2025 or 2030, and the technology is there to do that. Plug-in hybrid electric vehicles (PHEVs) can do even better, but care has to be taken to include emissions in generating the electricity used in determining comparative emissions. Population in the United States is growing by about 0.9% per year, so 50 mpg with the 20% population increase expected by 2030 reduces gasoline consumption by 40% for ordinary vehicles compared with today's population with today's vehicles, and the reductions are even greater with PHEVs. A 40% reduction in gasoline use by the light-vehicle fleet gives a 16% reduction in TPES. Investing more in advanced battery technology would be a winner. Continued large investment in hydrogen demonstration projects is a loser. Fuel cells need to go back to the laboratory to improve efficiency and catalysts.

The building sector is more complicated than transportation because of the fragmentation of the industry. It has many small producers, and is regulated by 50 separate states most of which have not set any energy standards for buildings. Better building codes, more stringent appliance standards, and cost-effective retrofit technologies for existing buildings are needed. Energy savings of 30% or more can be achieved in the buildings sector corresponding to another 10% of TPES. New buildings can do much better. (Don't double count here. About 80% of the energy use in buildings is electricity, so if we decarbonize electricity the efficiency improvements in the rest of the building sector only cut TPES by an additional 2% to 3%.) Increased funding in building technology would be a winner.

Substitution of emission-free fuels for fossil fuels was discussed in Chapters 12, 13, and 14. The two systems that today make up the largest amount of emission-free energy, nuclear, and big hydropower dams, run into strong opposition in the United States and EU from a small but very vocal group that I called the ultra-greens in the introduction.

For reasons I do not understand, they also do not seem to like carbon capture and storage (CCS) which, if successful, would allow the use of fossil fuels while making such use emission-free. The goal is to reduce emissions, and I am skeptical about achieving big emission reductions without nuclear and hydro power, but even with them would feel much more confident if CCS was made to work. Nuclear and hydro are winners. CCS received much support in a poorly designed program and needs to be rethought to have a chance to become a winner.

Chapter 14 reviewed biofuels. The US corn-based ethanol program should be terminated; it is a loser, but the federal government is very unlikely to do the right thing because Congress thinks it likely that votes will be lost in the corn-belt states if they end it. Brazilian sugarcane ethanol is a winner, and the EU's grain-based ethanol is somewhere in between. Phase-2 (cellulosic) and Phase-3 (advanced) are promising, but have not delivered yet. Almost all of the research money is going to short-term programs, which is a mistake. There will be no revolutionary advances as we are going now.

Cost is an issue in deploying emission-free energy sources. I confess that sometimes I do not tell my wife how much some new electronic thing costs, to avoid what can be a great deal of explaining depending on that cost. I have the feeling that the advocates of the renewables do not want the citizens who pay the bill to know how much they are paying. I like to think that we are all grown-ups and realize that a new technology is likely to be expensive at first, but will come down as more is deployed. Hiding the story is not right, so here is the story.

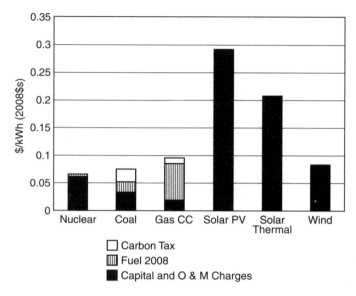

Fig. 15.2   Levelized cost comparison for electric power
generation with $100-per-ton tax on carbon (2008 fuel prices).
(*Source*: Prof. J. Weyant, Stanford University, Energy Modeling Forum)

Figure 15.2 shows the levelized cost at the power plant of various electricity sources in the United States (levelized costing takes the capital costs for a plant and spreads them uniformly over the life of the plant). Solar and wind costs are before any federal or state subsidies. Federal and state tax credits and rebates will reduce the charges on an electric bill by about 50% for solar and 30% for wind. Of course, the society is still paying the entire amount; the customer pays part and the taxpayer pays the rest. Costs are broken down into three components: capital plus operation and maintenance charges; fuel costs; and the added amount that would be imposed if there were a carbon emission tax of $100 per tonne of carbon. Coal is king without the emissions fee, and nuclear takes the crown with it. Gas CC is a high-efficiency combined cycle plant, and its costs are dominated by fuel prices. If there was much more gas, fuel costs would come down, but there will have to be a lot more gas to make electricity costs comparable to coal's even with a carbon tax.

The high cost of solar photovoltaic (PV) and solar thermal is caused by both the small fraction of the day that they generate power, and their high capital cost. The solar map in Chapter 13 showed that for a flat panel PV system the typical number of effective hours per year was only about 20% of the 8760 hour maximum. The capital cost of a PV system is about $5600 per kilowatt of capacity,[2] of the same order per kilowatt as a new nuclear plant. The nuclear plant runs 90% of the time and its power costs are so much less than the PV system mainly because the capital costs are spread over five times as many operating hours.

Solar thermal gets more sun time than PV because of its tracking system as shown in Figure 13.7, and its capital costs are somewhat lower as well. Solar costs will come down, but we will have to see how far down. Today, federal and state subsidies in California cut solar costs in half. Without those subsidies I doubt that we would see so much being deployed. It is too early to say how solar will come out. We know about solar thermal electric, but the laboratory may still produce revolutionary solar photovoltaic systems, without which photovoltaic is only an indulgence for rich countries. There is one exception: if there is a need for power far from any link to the grid, photovoltaic may be an affordable option.

Wind turbines have a lower capital cost than solar per kilowatt of capacity, and the wind in the United States typically blows about 30% of the time. Adding in tax credits makes wind very attractive as long as there is a backup for when the wind doesn't blow. Today, wind with a carbon tax would reach the magic cost level of grid parity compared with coal or gas and would be a winner (don't forget the new grid, but we need that anyway).

Changes in the ways we power our economy are not going to happen without carrots and sticks from government. For years we have been subsidizing various forms of energy at different rates. Recently, Management Information Services, Inc., in a study funded by the Nuclear Energy Institute, has analyzed where the US subsidies have gone.[3] Their results, covering the period from 1950 through 2006, show that tax policy, regulation costs, and federal R&D expenses make up over 80% of energy incentives. Oil received the most

---

[2] This capital cost estimate is from the EIA and is for plants of a few megawatts capacity. Home systems of a few kilowatt capacity in California cost about $8000 per kilowatt.

[3] www.nei.org/resourcesandstats/documentlibrary/newplants/whitepaper/ federal_Expenditures_for_Energy_development/

Table 15.1 *Federal R&D expenditures on three energy programs from 1992 through 2006 (2006 dollars)*

Program	Expenditures (billions of 2006 dollars)
Coal	6.08
Renewables	6.04
Nuclear energy	3.52

($335 billion) followed by natural gas ($100 billion), coal ($94 billion), hydro ($80 billion), nuclear ($65 billion), and renewables including geothermal ($52 billion). Tax breaks are what make oil and gas subsidies so large.

Of more interest to climate-change mitigation efforts are R&D expenditures since the 1992 Earth Summit in Rio de Janeiro. That was the time of the beginning of work under the United Nations Framework Convention on Climate Change. In the United States the bulk of R&D is carried out by the DOE, but other agencies including Agriculture, EPA, and NASA make small contributions to the programs. Table 15.1 shows data from the Management Information Services report in 2006 dollars. The numbers show clearly a shift in emphasis. What is missing from the Federal R&D portfolio is long-term R&D support which is where revolutionary change is most likely to come from.

In addition to federal support for R&D, a large amount is being spent in the form of the 30% tax credit for the installation of renewable energy sources. This credit has been very important in the deployment of wind and solar electrical systems. When the renewal of the tax credit was uncertain last year, construction of commercial-scale facilities of these types came to a near halt. It was extended just in time, but has expired three times, in 2000, 2002, and 2004, and each time new renewable plant construction nosedived.

The industrialized nations of the world can take the lead in controlling emissions both by cutting their own and by developing the technology for all to use. Cost-effective winners are desired. What technology will bring 50 years from now is unknown, but we should be starting with the winners in Table 15.2.

Table 15.2 *Energy winners and losers*

Winners	Losers	Maybes
Efficiency in all sectors	**Coal** without capture and storage	**Enhanced geothermal**
Coal with capture and storage	**Oil** to be replaced with electric or PHEV drive	**Solar thermal electric** needs cost reduction
Hydroelectric	**Corn ethanol**	**Solar photovoltaic** with subsidies (only for rich countries until costs decline)
Geothermal	**Hydrogen** for transportation	**Phase-2 and -3 biofuels**
Nuclear		**Ocean systems**
Wind, but needs tax credits now		
Gas as replacement for coal		
Solar heat and hot water		
Sugarcane ethanol		
Solar photovoltaic for off-grid applications only		
Advanced batteries for PHEVs		

Part III   Policy

16

US policy – new things, bad things, good things

16.1 INTRODUCTION

As of this writing (mid 2009), the United States does not yet have a national policy on emissions reductions. Until very recently it was the national champion emitter of greenhouse gases (China now has the title), a national policy was needed, none was forthcoming from Washington, and the states stepped into the breach. More than half of the states now have what are called Renewable Portfolio Standards (RPS). Some of the states' RPS are quite aggressive while others are mild. Some states already have significant fractions of their energy supply from emission-free sources, Maine, for example, with its large component of hydropower (45% of electricity in 2005). Regional collections of states have agreed on standards. What exists now is a patchwork of attempts to solve what is an international problem, and a national program is needed that places such a program in a world context. On the Federal stage, there has been a partisan divide, with the Democrats for action and the Republicans against. Among the states there was no such divide, and the regional compacts include states with Democratic and Republican governors. Emission reductions were never a partisan issue in the country, only in Washington.

The Congressional election of 2006 began to change the situation. New CAFE standards finally did become law in 2007, but there was no progress on broader greenhouse gas regulation. The election of 2008 changed things further, and greenhouse gas regulation is now on the front burner of the national stove. Drafts of model bills are circulating in the Congress, and the talk is of a national law before the end of 2009. The intent is to develop a control regime that will move the entire energy system toward low- or no-emission modes. That leaves the question of what to do about the state laws that have been enacted while Washington was paralyzed. I find that some are still useful, some are no longer useful, some were never useful, and some have become counterproductive. I hope that a national law brings a degree of uniformity to what is now jumble of different approaches. Here I will describe the approaches being discussed nationally and give my evaluation of some good and bad policies with my scorecard at the end.

16.2 REDUCING EMISSIONS ON A NATIONAL SCALE

There are two large-scale options under discussion. One is called Cap and Trade, while the other is called a carbon tax by most, but an emissions fee by me. I advocate the name change because, no matter how sensible a carbon tax might be, some troglodyte will mutter "no new taxes" and its political chances will go down. Both alternatives seek to move the country away from its present dependence on fossil fuels though in different ways. Cap and Trade has the political lead now.

Cap and Trade is proposed as an economically efficient way of reducing pollutants. It was first tried in the United States in an effective move to curtail acid rain by regulating and reducing sulfur dioxide (SO_2) emissions from coal-fired power plants (SO_2 mixes with water vapor in the atmosphere to form sulfuric acid; hence acid rain). Here is a simplified version of how such a system works. Each plant receives an allowance to emit a certain amount of SO_2 for some period of time, a year, for example. The total number of allowances is equal to the total emissions of all plants. For the next period the total allowance is cut by reducing the allowance for each plant by the same percentage as the total is reduced. A particular plant can switch to a cleaner fuel thereby emitting less, install equipment to catch the SO_2 before it gets out, buy extra allowances from other plants that can meet their required reduction at lower cost, or any combination of the three. The reductions continue through future periods until the goal is met. The system worked for SO_2 and the goal of the program was met faster and at lower cost than originally expected.

The attractiveness of Cap and Trade is that the total emissions are controlled and the emissions reductions are known in advance (if the system works). The trading option establishes a price for emissions and a market for their trade just like any other commodity, and allows the reductions to be made at the least cost. The problem is that the cost of such a program is unknown at the beginning, and will not be known until a market for allowances develops and stabilizes. As you might suspect there are other problems too, which I will get to after introducing the emissions fee idea.

An emissions fee is simply an amount that has to be paid to the government for every bit of CO_2e that is emitted. In Chapter 10 I showed how to set a fee for power plant emissions by charging what it would cost to capture and store the material, and came up with about \$45 per tonne of CO_2. There are other ways to arrive at a number but the theory is that in a market economy, if doing things the old way gets too expensive, you will move to newer and less costly ways. In the Introduction I called this tilting the playing field to make the market economy move in a desired direction by making it possible to increase profits by going in that desired direction. The attraction is that such a system is very simple to administer and is economically efficient. The costs are known in advance, but in contrast to Cap and Trade, the emissions reductions are not known in advance; you have to wait and see how fast the market moves things.

Where to apply the constraint is an issue in either case. In the jargon of the business, you can apply it upstream (as near the source as possible), or downstream (as near the user as possible). If it is coal, for example, upstream means at the mine while downstream means at the power plant. If it is oil, upstream means at the refinery while downstream means at the gas pump. This can get very complicated. If it is applied upstream in Cap and Trade, what do you do if the downstream user makes carbon capture and storage work? You lowered the amount of coal that can be sold at the mine, but the CCS-capable user needs the coal and gets rid of emissions some other way. It seems a bit easier to solve these problems with an emissions fee, but the problem exists for both solutions. No one has figured out how to make Cap and Trade work for transportation, and the European Union, which has a Cap and Trade system for electricity and some large industrial processes, does not even include transportation, saying it will rely on other measures to handle the problem.

Cap and Trade seems to me to have another problem: enriching the guilty and punishing the innocent. It is much easier for an inefficient producer to increase efficiency than it is for one who is already efficient. Here is a true-life example. Some years ago in one of California's periodic droughts, Stanford University, where I lived, began to talk about water rationing for residents. My wife is a conservationist so we already had low flush toilets, flow restrictors in the showers, drip irrigation, etc. Our neighbors had none of these. If we had been required to make a fixed percentage reduction, we could not have done so because we were already very efficient. Under Cap and Trade I would have had to buy credits from those profligate neighbors. Fortunately, the crisis passed before anything had to be done.

Translating this into the energy world, if you run an electricity generation plant using natural gas as the fuel, your efficiency for turning heat into electricity can range from 35% to nearly 60%. If you are required to reduce emissions by 10% it is much less costly to increase efficiency from 35% to 38.5% than it is to go from 50% to 55%. If you were already at 60% it would be impossible, for there is no more efficient power plant. You can only buy credits from the inefficient producer. Cap and Trade seems to create a new form of money in emission credits and most of it would appear to go to the least efficient producers. If the United States does introduce Cap and Trade, I hope the economists can figure out how to fix this problem. There is a very interesting collection of analyses by the Congressional Budget Office that look at the economic effects of Cap and Trade.[1] Their conclusions are that such a system does create a new form of money and if allowances are given away all the benefits go to the emitters and all the bills go to the tax payers. My economist friends tell me that auctioning emission permits rather than giving them away reduces the effect. As for me, if I could afford it, I would go out to buy the dirtiest coal-fired power plant that I could find, and plan on making my fortune by shutting it down and selling my permits.

There is some experience with Cap and Trade. I already mentioned the effective program to reduce SO_2 emissions in the United States. In the EU, Cap and Trade began in 2005 as a way to meet the European commitments to the Kyoto Protocol. The plan was to operate the system until 2007, and then modify it as experience indicated was necessary. In the first round, because of an over-allocation of free credits, the price for carbon emission permits collapsed and the program did not work well. It has been restarted with tighter controls over allocations and appears to be working better this time.

In the United States, industry has been advocating a hybrid plan which is basically Cap and Trade with an escape clause. They worry (not unreasonably) that the cost of emission permits may get to be too high and so want a system whereby the government will issue more permits as needed to keep the cost below some specified limit. Whatever is decided, it would be best if the experimental nature of

---

[1] http://www.cbo.gov/publications/bysubject.cfm?cat=18

the program was recognized and a specific time for re-evaluation was included in any legislation.

When it comes to a choice between the options, an emissions fee has one enormous advantage: it is very simple to administer. It should be imposed at the point where the payer has the option to change so as to reduce emissions. In electricity generation it should be at the power plant where the operator can switch to more efficient generators, introduce a CCS system, or even shut it down and build a wind farm. For transportation, it should go at the gas pump where the customer can chose to buy a car with better mileage.

When it comes to what to do with the revenue, the government can choose between many options including reducing income taxes, fixing social security, funding the development of more advanced emission-free energy, etc. That is a separate issue from the effectiveness of the program in reducing emissions. The fee will have to be adjusted over time to make sure that the desired reductions in emissions are achieved.

Administration is much more complex in Cap and Trade. It is easy to see what to do in electricity generation, but I can't see how to do it in transportation. Does each car owner get a gasoline ration card so that the allocation can be reduced over time, or do you keep vehicles out of the system and instead require changes at the car manufacturing point (better CAFE standards, for example)?

Setting aside the administrative problems, there are two advantages to Cap and Trade. The first is that the EU already has such a system, and integrating a US and EU system should have advantages for both. The second, which may in the long run be Cap and Trade's biggest advantage, is that it will work much better than a fee in non-market economies. In China, for example, the government owns the largest energy producers as well as a large piece of the coal industry. A fee just moves money from one pocket to another and gives no real national incentive to change behavior. Cap and Trade, with its assurances of emissions reductions, is more effective than a fee whenever the market and profits are not the economical drivers.

The United States is heading toward a version of Cap and Trade with all its administrative complexity at the time of this writing (mid 2009). Little attention seems to have been paid to European experience. Too many of the emission permits are to be given away and transportation is to be included. I would urge a hybrid system where we start with Cap and Trade for electricity generation, include an upper limit on carbon prices above which the government will issue more free permits, and treat the transportation system separately. We will probably not get it right the first time and should plan from the beginning to have a time for a re-evaluation as the EU did.

There is a second issue which is important – allowing emission reductions to be made in some other country where it may be less costly to do the job. After all, there is only one atmosphere, so in principle it does not matter where the reduction in emissions is made; only that it is made. These are called offsets in the Kyoto Protocol and there is no reason in principle that they could not be used to achieve national goals. They are used in practice to achieve the Kyoto goals of the EU, but have not worked well. I will come back to this issue in the next chapter.

16.3 BAD THINGS

In the name of saving us from global warming, there have been some bad policies introduced as well as some good ones. Greenness is very popular now, and things have been done in a rush, sometimes without looking carefully for unintended consequences. Here are a few examples.

Renewable portfolio standards

In all the talk about renewable energy, what seems to have been forgotten is that the enemy is greenhouse gas emissions and the hero to rescue us is low- or no-emission electricity, not just renewable energy. Several states in the name of greenhouse gas reduction have promulgated what are called renewable portfolio standards (RPS). Regrettably, the federal government is moving in the same direction. An RPS defines what fraction of electricity generated or used in the state must come from renewable sources. Renewable as now defined includes only solar, wind, geothermal, and small hydropower generation. Table 16.1 shows data from the EIA and EPA giving the percentage of electricity generated and the emissions produced by various fuels. It is misguided to exclude the largest producers of emission-free electricity, nuclear and big hydro, from the proposed standard.

Electricity costs at the power plant for the renewables were shown in Figure 15.2 before the inclusion of various tax credits and other subsidies. What is behind the RPS idea is to force the installation of renewable energy sources on a scale sufficiently large to begin to achieve the economies that go with large-scale production of the equipment. Wind power systems have come down in cost, but for the past few years costs have been flat even as the amount of wind energy installed goes up. While wind does get subsidies in the form of tax

Table 16.1 *Fraction of electricity generated and emissions from various fuels*

Fuel	Percentage of electricity*	Emission (Gt of CO_2e)
Coal	49%	1.9
Gas	22%	0.34
Nuclear plus big hydro	26%	0
Renewables	2%	0

*Rounding errors account for the difference from 100%.

credits, it is within sight of grid parity if an emission charge was added to coal and natural gas production of electricity. This is not true for solar electricity, and I think it unlikely that photovoltaic will approach grid parity without the introduction of some new kind of solar cell.

A greenhouse gas reduction standard (GRS) is more appropriate to reduce emissions. Under GRS, if carbon capture and storage was shown tomorrow to be workable and effective, it would qualify, but with an RPS a good way to reduce emissions while still using fossil fuels would not be allowed. If there were to be enough gas to replace coal, US emissions would go down by 1200 million tonnes per year, something that should be desired, but certainly is not encouraged by an RPS. Nuclear power would reduce emissions even further.

I have said we have to move away from fossil eventually, but greenhouse gas reduction is the goal, and using any fuel now in an emission-free fashion meets that goal and gives time to develop better longer-term solutions. We should be setting GRS not just RPS, unless there is more to the goal than what is announced.

Balkanization – cars and California

California has always been a leader in environmental regulations. From smog control to efficiency standards for household appliances, it has tended to lead the nation. In 2002 California set out to regulate greenhouse gas emissions for cars by requiring that the light-vehicle fleet sold there be able to average at least 35 miles per gallon of gas. The federal standard had stayed at 25 mpg (average of cars and light trucks) since 1985, and the Bush administration showed no signs of action on any front having to do with greenhouse gas emissions. The state has authority under the Clean Air Act to set more stringent air quality standards than the federal ones, from the days when California had smog control standards and the federal government did not. The state tried to use its authority under the Clean Air Act to set its own limits.

From 2002 on, the issue of the state's authority to so regulate moved through the courts. The auto industry sued to block California's action while others sued to force the EPA to set emissions standards for vehicles. All the legal maneuvers came to an end with a Supreme Court decision in April of 2007 that said that the Clean Air Act gave the EPA the right to regulate, that the EPA could decide not to regulate, and the grounds on which the EPA had decided not to regulate were not good enough. The final act was an EPA decision not to regulate and to refuse to grant California the waiver required to let it set its own standards.

The new administration has raised the curtain on the next act of the drama. President Obama ordered the EPA to review its decision refusing California a waiver. The waiver will soon be forthcoming, and what will California do then? I expect that it will impose its 35 mpg requirement on new cars sold in California beginning in 2016. It sounds like a victory for the environment and for the states (several are expected to follow California), but if California does set its own standards, it will be a defeat for the country. What was a bold move when California started out in 2002 would be an unwise move now. There has been one big change since all this started in 2002. There is a new federal standard of 35 mpg beginning in 2020 (the EPA has asked for public comment on moving the national standard's effective date to 2016).

I don't see the sense at this time of saddling the beleaguered auto industry with a local requirement to do four years earlier what they are already required to do for the entire country. I have no doubt that they can make a 2016 deadline for some of their models, but given their present financial situation, I do doubt that they can do it for all of their models. There is not enough to be gained by advancing the due date for 35 mpg in a few states. It would be much better in the long run to be working toward a 50 mpg standard in 2025 or 2030.

Low carbon fuel standards (LCFS)

According to the Office of Science and Technology Policy's website (May 2009), the Obama administration wants to implement a low carbon fuels standard (LCFS) like that recently introduced in California. This is a bad idea because it introduces a complicated program with relatively small gains in emissions while making the changes that can have much bigger effects harder to do.

In the California version of a LCFS the fuel used in transportation has to be modified so that for the same energy content, the carbon content (and supposedly thereby emissions) is reduced by 10% by 2020 compared with today. Emissions from all stages in fuel preparation are counted, including those from getting it out of the ground whether it be oil or corn, transportation, changes in land use, etc., as well as from burning the fuel itself. It sounds good, but fuel is only part of the story. What counts is what comes out of the tailpipe, or what I call emissions per vehicle mile traveled (EVMT) which depends on both the fuel and the efficiency with which it is used.

Those for a LCFS argue that in transportation it is the energy in a fuel that moves a vehicle and we should reduce the greenhouse gas emissions associated with using that energy counting all the emissions made in producing the fuel. Counting everything in the fuel chain is good, but the fundamental premise is wrong. My wife's Prius has half the EVMT of a conventional car of the same size using today's gasoline. A vehicle with a diesel engine is about 20% more efficient than a gasoline engine because it runs at higher compression and temperatures, so a diesel will reduce EVMT by 20%, all other things being equal. The new national CAFE standard of 35 mpg will lower the average EVMT by 29%. Plug-in electric vehicles have an EVMT that depends on how electricity is produced, and averaged over the county's electric system is already better than required by the LCFS. All of these are ready for large-scale implementation now and reduce emissions far more than an LCFS. Efficiency improvement can dwarf the emissions reduction coming from an LCFS, so why not focus on the big rewards rather than on a very complicated little thing?

The advocates of the LCFS argue that there are some fuels that make emissions worse and we should force them out. An example is corn ethanol produced where there is a lot of fossil fuel used to make the electricity that runs the factories. Everyone knows that, and I am sure that the politicians who mandated today's ethanol subsidies to agribusiness will find a way to cancel anything attempted on a national scale to do away with the subsidies. LCFS fans also don't like oil from the Canadian tar sands where it takes from 15% to 30% of the energy content of the fuel to get it out of the ground and through the refinery.

Some people that I respect are enthused about an LCFS, but they have never been able to make clear to me why California or the nation should be doing it now. Having read the 367-page volume defining the California LCFS, I am a lot clearer on just how complicated it will be to implement, but no clearer on the justification for the program. Cleaning up emissions from the transportation sector is a very big job. We should focus first on the things that can have the largest effects with the fewest complications, and the recent study by the American

Physical Society shows just how big the impact of improved efficiency can be.

California, for many years, has been a leader in energy efficiency and emissions control, but I hope that the other states that look to California for a lead will pass this one by. We should focus on the result we want and that is reduced emission per vehicle mile traveled, not what is in the fuel. Reduced emissions can be better done with better cars like hybrids or plug-in hybrids, or with further improvements in CAFE standards. The LCFS only makes it harder to get to where we want to be. After we get the big and easy things done it may be worthwhile to come back to the fuel.

16.4 GOOD THINGS

Crossing the Valley of Death

Basic research is often said to be the foundation of innovation, but there is a long road from the laboratory to the market place, and that road has become more difficult to traverse in the past several decades. The great industrial research laboratories like the fabled Bell Labs that produced the transistor and a number of Nobel Prizes no longer do much long-term research and development. Industry and the financial markets have changed and the focus is now on the quarterly financial statement, which makes it difficult for managers to justify the investment in long-term efforts that will produce no income for many years if their competitors do not do so as well. Such work puts more money on the expense side of the ledger and less on the income side, and no one wants to seem to be doing less well than their rivals.

Support of long-term research has become the province of governments, and industry tends in most cases to focus on work that can lead to a marketable product in no more than five or so years. But, on what I called the road from the laboratory to the marketplace, there is often a long development stretch which seeks to see if the laboratory invention can really make something useful. Traversing this has come to be called crossing the Valley of Death in the innovation chain. This part of the road from discovery to the visibility of a broad application needs funding, and the research portfolio of governments needs to reflect this reality. It often doesn't, and since industry no longer routinely supports this kind of work, this stretch of the road is often the graveyard of good ideas.[2]

[2] The National Academy of Sciences has a website www.beyonddiscovery.org that gives the history of important developments from their first discovery to their application.

Part of the problem may be a fear of failure and the criticism that goes with it on the part of government officials. They might take a leaf from the book of the venture capitalists in Silicon Valley. They expect that of every ten investments they make, five or six will fail, three or four will limp along, and one will be a major success. That one makes up for the other nine. Federally funded R&D tends to be risk averse, perhaps because of fear of criticism from those who do not understand the game. Some years ago the US Congress asked the National Academy of Sciences to evaluate the DOE's energy R&D program [47]. The report said that not only was it worth it but that the societal benefits of a few of the programs more than paid for the entire portfolio. The venture capitalists would nod in agreement.

The stimulus bill recently passed by the US congress has in it an attempt to solve the problem. It contains money for a new part of the DOE called E-ARPA or the Energy Advanced Research Projects Agency. It is modeled after the Defense Department's ARPA which has been a great success, among other things funding the development of the first large-scale integrated circuits that are at the heart of all of our computers, and developing the first stages of the Internet that business and industry rely on today. E-ARPA is new and will need time to get organized and start operations. If it does well, crossing the valley will become much easier. It is worth watching and if successful, emulating.

Demand side management

This is a good program, also a California invention that has spread to some, but not all, of the states. What the California Public Utilities Commission did was to invent a method to allow utilities to make money by getting customers to use less of their product. Before DSM, regulated utilities could only earn a fixed percentage return on sales, so to make more money they had to increase energy use. The CA-PUC did not tell utilities how to do things, only that they would be allowed to earn more if they succeeded, and succeed they did. DSM should be expanded to more places.

The result is interesting. California electricity costs are high per kilowatt hour, but the use of electricity is down so that customers end up paying less in total. This is the kind of thing California's regulators should have done instead of the Low Carbon Fuel Standard, and I have never understood why the state does not use its own invention to decrease emissions from autos.

Table 16.2 *Policy scorecard*

Winners	Losers
Cap and Trade for power and some industries	**Cap and Trade** for transportation
Emissions fees	**Renewable portfolio standards**
Greenhouse gas reduction standards	**State mile per gallon standards**
Federal mile per gallon standards	**Low carbon fuel standards**
Demand side management	
Long-term development funding	

Table 16.2 is my scorecard for the policy options discussed here. It will surely displease some. Cap and Trade works in some sectors of the economy and not in others. Why the United States seems to be trying to use it in an area where no one has been able to make it effective is a mystery. Emissions fees work in all sectors. Greenhouse gas reduction standards are preferable to renewable portfolio standards. After all, what we are all trying to do is reduce emissions and that should be done in the lowest cost manner, not just by what some consider the greenest route. Federal mileage standards are better for the United States than state standards. As long as they are tight enough, it would be far better to have uniform standards across the United States, and if Europe introduces them too, across the entire European Union. Demand side management tilts the playing field so that industry can make more money by doing the right thing rather than the wrong thing, a method simpler than trying to make detailed regulations. Long-term development funding can be the source of fuel to get across the Valley of Death. Low carbon fuel standards sound good, but seem to get in the way of efficiency measures in transportation.

17

World policy actions

17.1 INTRODUCTION

All the nations of the world share one atmosphere. What goes into it affects all, and the consequences of climate change will fall on all. Depending on your perspective and your experience with international agreements, you can be either impressed or disillusioned about the international response to the need to mitigate global warning. It was only in 1992 at the Rio Earth Summit that the nations of the world agreed there was a problem. After that, with remarkable speed,. the Kyoto Protocol was produced in 1997, and entered into force in 2005 when industrialized nations accounting for at least 55% of 1990 emission signed on.

A mere 13 years from recognition to action is regarded as fast by those with experience with the UN organization, or slow by those focused on the urgency of the problem. However you regard it, Kyoto is the basis for action now, but it expires in 2012 and has to be replaced with a new and necessarily better Protocol that brings in all the nations that were left out of the action agenda last time.

The United States played an important role in designing the Kyoto Protocol, but never ratified it. Cap and Trade is an example of a US proposal that was viewed with suspicion at the beginning of negotiations, but became the mechanism favored by most for reducing emissions. There are other US inventions as well. However, the issue at home in the United States before, during, and after the Kyoto meeting was the role of the large developing countries. The US Senate in a resolution in early 1997 stated clearly that they would not ratify any treaty that did not include some binding commitments on the part of the developing countries. There were none included, and President Clinton did not send the Protocol to the Senate for ratification since he knew that it would lose. On taking office President G. W. Bush said he would not send it on since it would not work if the developing counties made no commitments of their own.

In a piece published in the *New York Times* of April 17, 2001, I said that I agreed with President Bush on the Protocol. This briefly made me a darling of the deniers who failed to notice the part that said the important issue was what the United States would propose to improve it. I was reminded of a line in Gilbert and Sullivan's *The Mikado* where Pooh-Bah sings, "And I am right and you are right and everything is quite correct."

President Bush's position was based on projections that in a few decades the developing world would be emitting more greenhouse gases than the industrialized world. If the developing world went on with business as usual, and the industrialized world reduced its CO_2 emissions to zero, by 2050 we would still be worse off than we were then. He saw the Kyoto Protocol as fatally flawed because it did not commit the developing nations, and he was right.

China (India, too) had a huge population and a low standard of living. Their only hope to improve the lot of their people was economic growth, and growing they were, and still are. They said, "You first." They said that they did not cause the problem, they were poor and needed to grow their economy to improve their standard of living, and they were right.

The Europeans said, "Let's stick to the Kyoto Protocol and bring the developing world in later." There may have been some hypocrisy in their position. Europe had a smaller economic growth spurt during the 1990s than did the United States, and it was easier to meet the Kyoto goals. Nonetheless, Europe said it was willing to do hard things to meet the CO_2 targets, the problem was urgent, and the industrialized world should begin to do something immediately, and they were right.

Among all this "rightness" the United States was the least right because the others were willing to do something while it did nothing and proposed nothing, and the Protocol has been dead in the United States throughout the entire Bush administration. The Obama administration is committed to action, and we will see what happens with the forthcoming renewal of the Protocol, which I will call Kyoto-2.

17.2 KYOTO-1: THE PROTOCOL OF 1997

The Protocol was regarded by all its signatories as a first attempt at control of greenhouse gas emissions.[1] It was recognized at the

---

[1] http://unfccc.int/resource/docs/convkp/kpeng.html

beginning to be an imperfect instrument and so had a finite duration (until 2012); it was expected to be replaced by a better instrument where experience would be a teacher pointing to a better way. The nations of the world were divided into two classes, the industrialized and the developing. The industrialized nations were to take the lead and only they, as listed in Annex B of the Protocol, were committed to binding emissions targets. The main Annex B nations and their targets for reductions relative to their emissions in the base year of 1990 are the European Union (–8%), the United States (–7%), Japan (–6%), and the Russian Federation (0%), plus a few others including most of the Eastern European members of the former Soviet block. Though the goals are called binding commitments, there are no sanctions specified in the Protocol for failing to meet the goals.

The United States has not ratified the Protocol though all the other Annex B countries finally have (the last was Australia in November 2008). Even if all the goals specified in the Protocol were achieved on schedule, the continued increase in emission would change very little. Figure 17.1 is a repeat of Figure 6.2 and shows the expected increase in TPES. The increase is dominated by the growth of the developing counties and so merely setting back the emissions of the industrialized countries to 5% below 1990 could only have a tiny effect. A close look at

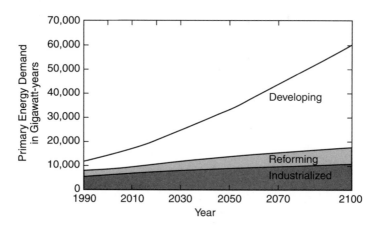

Fig. 17.1   IIASA projection of future primary energy in their high growth scenario. IIASA projections show that energy demand in the twenty-first century is dominated by the growth of the developing nations. (*Source*: International Institute of Applied Systems Analysis and World Energy Council Global Energy Perspectives long-range projection; choose different assumptions here: http://www.iiasa.ac.at /cgi-bin/ecs/book_dyn/bookcnt.py)

the figure shows that the goals would only set the increase path back about 4 or 5 years because only the industrialized nations are committed to reductions under Kyoto-1. Instead of TPES ending at 60 000 gigawatt-years in 2100 it would be reduced to about 57 000 gigawatt-years. The hope was and is that low- or even no-emission technologies would be developed that could be used in the developing countries as well as by the industrialized nations to meet their commitments.

The Protocol also has a mechanism to allow credit toward an Annex B country's goal to be gotten by reducing emissions in a developing country. This is called the Clean Development Mechanism (CDM). There is a requirement for something called "additionality" for a project to get credit under the CDM. This means that the project is something the developing country would not have done anyway. There is a complicated formal process for a project to get designated as a CDM project, and as you can guess a cohort of consultants has sprung up who are good at the bureaucracy. I have discussed the CDM with several experts and the consensus is that it has been a failure.[2] Additionality has been too easy to get accepted, and clean up of some of the more powerful, exotic, and easily removed greenhouse gases has earned an emitter in a developing country far more money under the CDM than it would have cost that country to clean up directly.

From the perspective of reduction in emissions, the Kyoto Protocol of 1997 has done very little. It has tested out some mechanisms, good and bad, and has given time for some critical thinking about what is next. Most important, the explosive growth of the economies of China and India, and the emissions that have gone with it, have convinced most that the developing countries have to be brought into the system in the successor to the 1997 Protocol.

The attitude of the public has changed greatly since 1997. Back then, few paid attention. Today the public is engaged and there is a broad consensus that the United States has to join the effort to mitigate global warming. The sentiment in the US Congress is such that it is highly likely that some sort of emission control law will be passed this year or next if the economy does not get much worse. The political sentiment is such that it will probably be a Cap and Trade program. There will be much argument about the details of emission allowances and where to impose limits, and it is to be hoped that a good law will emerge from all the jockeying for advantage. It is not likely that the United States will sign on to Kyoto-1, since it is far too late to get anything significant done by 2012.

---

[2] An interesting, though somewhat technical analysis is given in Ref. [48].

17.3 KYOTO-2

What I have been calling Kyoto-2 is to take effect in 2012 when the original Protocol expires. A meeting of all the signatories to the UN Framework Convention on Climate Change is to take place in Copenhagen in December of 2009 to try to come to agreement on the next steps. The big questions will be the period covered by the next Protocol, the greenhouse gas reduction targets, and the role of the developing nations, particularly China and India which are the two with the most rapidly increasing emissions levels. I will simply assume the duration is to 2030 and the target is the level required then if the world is to be on the emissions trajectory required to stabilize the atmosphere at 550 ppm (twice the preindustrial level). There will be a great deal of discussion about these two things, but the most noise will be about the role of the developing countries. The tale told before Kyoto-1 (the industrialized countries caused the problem and they should bear the burden of cleaning things up) will no longer do. By the year 2100 the business-as-usual trajectory for TPES will have the developing countries contributing nearly as much greenhouse gas to the atmosphere in this century as the industrialized ones will have contributed in the 300 years from 1800 to 2100. We are all in this together and if there is no agreement on participation by these countries we cannot fix this problem.

I picked the year 2030 for my target because that is about the year that most analyses give as the time when emissions have to peak if we are to stabilize at 550 ppm. Some analyses say 2020, some say 2050, but most show an emissions trajectory that is fairly flat in the 2020 to 2050 time frame.

Figure 17.2 shows the International Energy Agency's estimates of CO_2 emissions from the energy sector for their reference scenario (similar to business as usual, but not exactly the same) and for stabilizing the atmosphere at 550 ppm of CO_2. By 2030 the reduction required is about 8 gigatonnes (Gt) of CO_2 below the reference. *This is not going to happen without some degree of participation by all the nations of the world.*

The IEA estimates that in 2030 in the reference case the OECD members[3] (mostly what I have called the industrialized countries)

[3] The 30 member countries of OECD are: Australia, Austria, Belgium, Canada, Czech Republic, Denmark, Finland, France, Germany, Greece, Hungary, Iceland, Ireland, Italy, Japan, Korea, Luxembourg, Mexico, the Netherlands, New Zealand, Norway, Poland, Portugal, Slovak Republic, Spain, Sweden, Switzerland, Turkey, United Kingdom, United States.

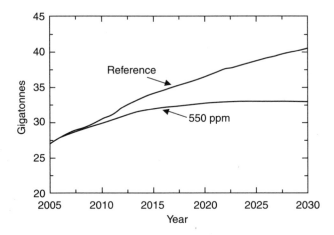

Fig. 17.2   IEA emission scenarios. IEA "business-as-usual" projection for
emissions and the world emissions versus time allowed in their model
if the atmosphere is eventually to be stabilized at 550 ppm. (Adapted
from © OECD/IEA, *World Energy Outlook* 2002, p. 338, Fig. 12.1)

would emit about 13 Gt of CO_2 while the rest of the world emits 28 Gt.
To achieve the needed 8 Gt emission reduction needed for the 550 ppm
scenario would need the OECD members alone, including the United
States, to reduce emissions by 60% of projected emissions. I don't
think they can or will do it while the rest of the world does nothing.
There is no way the Kyoto Protocol's Annex B countries (the only ones
required to meet mandatory goals under Kyoto-1), roughly the OECD
members plus the Russian Federation, will be willing to meet it on
their own either. According to the OECD analysis, 75% of the increase
in emissions between now and 2030 will come from China, India, and
the Middle East, and a way has to be found to involve them.

Table 17.1 shows data for the world and the top ten emitter
countries (I have grouped the EU as one entity because they act as one
in emission control). The data are from 2005 (the most rapid growth in
emissions since has been in China which has now passed the United
States as the largest emitter).

There is a strong correlation between GDP and total emissions.
The world average is an emission of about one-half tonne of CO_2 per
$1000 of GDP. The United States is about at the average, China is above
it, the EU is below, but the range of values is small. But, there is a
wide spread in per capita income and this will be the source of the big
problem confronting the designers on Kyoto-2. (Emission per capita is
mainly a reflection of per capita income.) The Annex B list of countries

Table 17.1 *Greenhouse gas emission indicators*

Region or country	Population (millions)	CO_2 emissions (million tonnes)	GDP (PPP) billion (2000$)	GDP (PPP) per capita	CO_2 tonnes per capita
World	6432	27,136	54,618	8492	4.2
United States	297	5817	10,996	37,063	19.6
China	1311	5101	8057	6146	3.9
EU	492	4275	11,608	23,605	11.8
Russia	143	1544	1381	9648	10.8
Japan	128	1214	3474	27,190	9.5
India	1095	1147	3362	3072	1.1
Korea	48	449	958	19,837	9.3
South Africa	47	330	463	9884	7.0
Brazil	186	329	1393	7475	1.8
Saudi Arabia	23	320	323	13,977	13.8

Population, emissions, GDP (PPP), GDP per capita, and emissions per capita show the wide variation in per capita income among the world's top ten emitters of greenhouse gases according to 2005 data. (*Source*: IEA *Key World Energy Statistics* 2007)

in Kyoto-1 committed to action excluded the developing countries from any requirement to reduce or even to slow the rate of increase of emissions. In the negotiation over Kyoto-2 the claim will be made that the developing countries are too poor and should be allowed to keep on building their economies while others take care of global warming problems. I hope I have convinced the reader that no stabilization scheme can succeed without including them. I see two major issues that will confront the negotiators.

- What are the standards that determine if a country has some commitment to action?
- What are the rules that relate economic development to the size of the action that a particular country should take?

Much earlier in this book I observed that there are 1.6 billion people with no access to commercial energy, and giving them minimal energy from even the worst of the coal-fired power plants would only increase emissions by about 1%. These people live in the poorest countries, and it should be possible to set some threshold value of per capita national GDP before participation becomes the normal

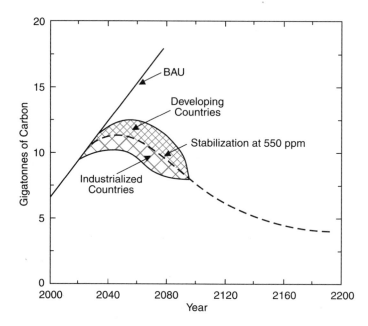

Fig. 17.3 A global model for stabilization. The dashed line is the
allowed emissions to stabilize the atmosphere. A world agreement
might, for example, allow the developing nations to run above the
dashed line while the industrialized countries run below. Where a given
region or country lies should depend on both its emissions and its state
of development. Eventually all will have to agree to a common measure.

expectation. If this is the mode adopted for exclusion from action, you
can bet that those countries part way up the development ladder will
argue to set this number as high as possible.

For the second part of the question that relates income to
action, I have a vague picture in my mind that is shown pictorially in
Figure 17.3.

The dashed line shows a stabilization trajectory for allowed
emission versus time that will stabilize the atmosphere at 550 ppm. In
my picture the industrialized countries have to reduce their emissions
faster than the developing countries and their commitments to reduction lie in the band below the dashed line, while the developing countries' commitments are in the band above. How to set the target for
individual countries will be a difficult negotiation. Here, I think the
European Union can be a model for all. In Kyoto-1 the EU commitment
is as a bloc and they determine the individual allocations within the
bloc. There are now 27 members of the EU, varying widely in per capita
income. The poorest (2005 data) are Latvia and Poland with per capita

incomes of around $8000 each (PPP). The richest are Luxembourg ($56 000) and Ireland ($34 000). The EU through its own process determines who does what and set goals for the individual members. Perhaps a regional approach might work in other areas as well.

I hope these negotiations come out well. I do not think the Copenhagen meeting in December 2009 is a good forum for all the negotiations because there are too many voices there. It would have been far better if a smaller group of countries that included the largest emitters in both the industrialized and the developing nations had come to some sort of preliminary agreement on general principles. President Bush started such discussions but I do not think they got very far before he left office. President Obama is continuing them. I hope we will see some sensible policies and programs emerge, but time is short. It would be better to take another year to get an agreement than arbitrarily to require one in 2009. Designing a system that is in the economic interest of the developing world as well as the environmental interests of all of the nations of the world is going to be no easy task.

As of now (August 2009) I see little progress toward a new agreement. In the United States, legislation to limit greenhouse gas emission has been loaded with special interest exceptions to the point where it doesn't seem to limit much. There are signs that no national legislation will be completed this year, although individual states continue to develop innovative programs. If there is no national US action, there is no way that the developing nations are going to agree to serious actions of their own. After all, the United States emits 19 tonnes of greenhouse gases per person while China and India emit 4 and 1 tonnes, respectively.

Perhaps a different approach might be considered. Now the focus is on legally binding commitments, but governments are very reluctant to sign such agreements when they do not know how to accomplish the agreed goals. It is much easier to get all to act when each sees an advantage. An example is the Montreal Protocol to phase out the fluorocarbons used in air conditioners that were destroying the ozone layer in the atmosphere that protects us all from dangerous ultraviolet radiation. A substitute was found and getting an agreement to phase out the bad for the better was not difficult.

Putting on the brakes to slow and eventually stop global warming is a much more difficult job since its causes are so tied to economic activities. However, certain actions can be to every nation's advantage, developed or developing. An example is energy efficiency, where less energy with fewer emissions is used to do a particular job. China continues to expand its use of coal for electricity, but does so mostly with the most efficient power plants. A broad agreement to use the most efficient technologies in all areas of economic activity will be much easier to negotiate than one with binding commitments that no one knows how to reach.

Since I have become involved in energy and proliferation issues I have learned one new thing: politics – particularly international politics – is much harder than physics.

18

Coda

The next few years will be important in determining the course of world efforts to control climate change. Our eyes and ears are constantly bombarded with claims and counterclaims that I said in the Introduction included the sensible, the senseless, and the self-serving. My aim has been to tell the technical side of the story in an honest fashion at a level for the general public without oversimplifying or hiding the consequences of the choices that must be made. I hope that armed with some facts the reader can distinguish between my three Ss.

Many say that in democracies the people are rarely willing to make hard choices unless they are frightened of the consequences of not making them. Perhaps I am too much the romantic, but I believe that they are willing when the problem is clear and the consequences of action or inaction are clear. I hope you will take a few things away from this book:

- The greenhouse effect is real;
- We are changing the atmosphere and the world is heating up;
- The science is still evolving and how bad things might become is still uncertain;
- Inaction is certain to have serious consequences;
- The longer we delay starting to deal with climate change, the harder dealing with the problem will be;
- The problem is emissions of greenhouse gases and the goal is to reduce them: the world does not have to run only on windmills and solar cells;
- We can mitigate the damage, but have to act on a worldwide scale;
- The richer countries will have to develop the technologies that all can use;

- It will be hard to develop sensible national policies and even harder to develop sensible international ones, but we must try to do so.

If we do nothing, it is our grandchildren who will begin to see the worst effects of climate change, and it is our grandchildren for whom we should all be working.

References

[1] G. Hardin, The Tragedy of the Commons, *Science*, **162**, 1243 (1968)

[2] J. P. Peixoto & A. H. Oort, *Physics of Climate* (Springer, 1992)

[3] C. D. Keeling, The concentration and isotopic abundances of carbon dioxide in the atmosphere, *Tellus* **12**, 200 (June 1960)

[4] A. Maddison, *The World Economy, A Millennial Perspective* (OECD Development Center, 2003)

[5] *Climate Change Science, An Analysis of Some Key Questions* (National Academy Press, 2001)

[6] C. L. Sabine, R. A. Feely, N. Gruber *et al.*, The ocean sink for anthropogenic CO_2, *Science* **305**, 369 (2004)

[7] IPCC, *Climate Change 2007: Synthesis Report*, http://www.ipcc.ch/pdf/assessment-report/ar4/syr/ar4_syr.pdf

[8] D. Canfield *et al.*, *Science* **315**, 92–95 (2007), www.sciencemag.org

[9] G. H. Roe & M. B. Baker, *Science* **318**, 629 (2007)

[10] *World Energy Perspectives, A Joint IIASA–WEC Study* (Cambridge University Press, 1998), http://www.iiasa.ac.at/cgi-bin/ecs/book_dyn/bookcnt.py

[11] *Key World Energy Statistics* (International Energy Agency, 2008), www.iea.org

[12] *Inventory of US Greenhouse Gas Emissions and Sinks 1990–2006*, USEPA report # 430-R-08–005 (US Environmental Protection Agency, 2008). This report has detailed data on emissions from all major sources.

[13] S. Pacala & R. Socolow, *Science* **305**, 968 (2004)

[14] *Annual Energy Review 2007*, DOE/EIA-0384(DOE/EIA, June 2008), www.eia.doe.gov

[15] *Resources to Reserves* (International Energy Agency, 2005). This reference has detailed data for both oil and gas and is an excellent source of information on extraction technologies as well. http://www.iea.org/textbase/nppdf/free/2005/oil_gas.pdf

[16] P. J. Meier, *Life-Cycle Assessment of Electricity Generation Systems with Applications for Climate Change Policy Analysis*, PhD dissertation, University of Wisconsin (2002)

[17] S. White, *Emissions from Helium-3, Fission and Wind Electrical Power Plants*, PhD dissertation, University of Wisconsin (1998)

[18] M. K. Mann & P. L. Spath, *Life Cycle Assessment of a Biomass Gasification Combined-Cycle System* (1997), www.nrel.gov/docs/legosti/fy98/23076.pdf

[19] W. Krewitt, F. Hurley, A. Trukenmüller & R. Friedrich, Health risks of energy systems, *Risk Analysis* **18**, 377 (1998). The paper contains many more health-based metrics.

219

[20] *Energy Efficiency Indicators for Public Electricity Production from Fossil Fuel* (OECD/ IEA, July 2008), http://www.iea.org/textbase/papers/2008/En_Efficiency_ Indicators.pdf

[21] B. Metz, J.C. Abanades, M. Akai *et al.*, *IPCC Special Report on Carbon Capture and Storage* (Cambridge University Press, 2005), www.ipcc.ch/pdf/special-reports/srccs/srccs_summaryforpolicymakers.pdf

[22] G. W. Kling, M. A. Clark, H. R. Compton *et al.*, The 1986 Lake Nyos gas disaster in Cameroon, West-Africa, *Science*, **236**, 169 (1987)

[23] S. C. Davis & S. W. Diegel, *Transportation Energy Data Book*, 26th edition, ORNL-6978 (Oak Ridge National Laboratory, 2006), http://cta.ornl.gov/data/ index.shtml

[24] American Physical Society Energy Efficiency Study Group, *Energy Future: Think Efficiency* (APS, 2008)

[25] D. Gordon, D. Greene, M. Ross & T. Wenzel, *Sipping Fuel and Saving Lives* (ICCT, 2007), www.theicct.org/documents/ICCT_Sipping FuelFull_2007. pdf]

[26] D. Santini & A. Vyas, "How to Use Life-cycle Analysis Comparisons of PHEVs to Competing Powertrains," presented at 8th International Advanced Automotive Battery and Ultracapacitor Conference, Tampa, Florida, May 12–16, 2008

[27] D. Santini & A. Vyas, "More Complications in Estimation of Oil Savings via Electrification of Light-duty Vehicles," presented at PLUG-IN 2008 Conference, San Jose, CA, July 2008

[28] A. Vyas & D. Santini, "Use of National Surveys for Estimating 'Full' PHEV Potential for Oil-use Reduction," presented at PLUG-IN 2008 Conference, San Jose, CA, July 2008

[29] *Reducing Greenhouse Gas Emissions: How Much at What Costs* (McKinsey and Company, 2007), www.mckinsey.com/cleintservices/ccsi/greenhousegas.asp

[30] R. Brown, S. Borgeson, J. Koomey & P. Biermayer, *Building-Sector Energy Efficiency Potential Based on the Clean Energy Futures Study*, LBNL-1096E. (Lawrence Berkeley National Laboratory, September 2008), http://enduse. lbl.gov/info/LBNL-1096E.pdf

[31] *The Economics of Nuclear Power* (World Nuclear Association, July 2008), http:// www.world-nuclear.org/info/inf02.html

[32] M. Bunn, B. van der Zwaan, J. P. Holdren & S. Fetter, The economics of reprocessing versus direct disposal of spent nuclear fuel, *Nuclear Technology*, **150**, 209 (2005)

[33] N. N. Taleb, *The Black Swan: The Impact of the Highly Improbable* (Random House, 2007)

[34] *Get Your Power from the Sun; A Consumer's Guide*, www1.eere.energy.gov/solar/ pdfs/35297.pdf

[35] REN 21 Group, *Renewables 2007 Global Status Report*, http://www.ren21.net/ globalstatusreport/default.asp

[36] *The Future of Geothermal Energy*, http://www.inl.gov/technicalpublications/ Documents/3589644.pdf

[37] *An Evaluation of Enhanced Geothermal Systems Technology*, http://www1.eere. energy.gov/geothermal/pdfs/evaluation_egs_tech_2008.pdf

[38] *Feasibility Assessment of the Water Energy Resources of the United States for New Low Power and Small Hydro Classes of Hydroelectric Plants*, DOE-ID-11263 January 2006, http://hydropower.id.doe.gov/resourceassessment/pdfs/main_report_ appendix_a_final.pdf

[39] OECD, *OECD Factbook 2008*, http://puck.sourceoecd.org/vl=2931846/cl=16/ nw=1/rpsv/factbook/050106.htm

[40] G. Boyle (editor), *Renewable Energy*, 2nd edition (Oxford University Press, 2004)

[41] P. E. McGovern, *Ancient Wine: The Search for the Origins of Viniculture* (Princeton University Press, 2003). The book traces wine making back some 7000 years.

[42] A. E. Farrell, R. J. Plevin, B. T. Turner *et al.*, Ethanol can contribute to energy and environmental goals, *Science*, **311**, 506 (2006)

[43] *Water Implications of Biofuels Production in the United States* (National Academies Press, 2008); www.nap.edu/catalog/12039/html

[44] C. B. Field, J. E. Campbell & D. B. Lobell, Biomass energy: the scale of the potential resource, *Trends in Ecology and Evolution*, **23**, 66 (2007)

[45] R. D. Perlock, L. L. Wright, A. F. Turhollow *et al.*, *Biomass as Feedstock for a Bioenergy and Bioprod ucts Industry*, DOE/GO-102995–2135 (April 2005), http://feedstockreview.ornl.gov/pdf/billion_ton_vision.pdf

[46] T. Searchinger, R. Heimlich, R. A. Houghton *et al.*, Use of croplands for biofuels increases greenhouse gases through emissions from land use change, *Science*, **319**, 1238 (2008)

[47] *Energy Research at DOE: Was It Worth It?* (National Academy Press, 2001)

[48] M. Wara, Measuring the Clean Development Mechanism's performance and potential, *UCLA Law Review* **55**, Issue 6 (2008), http://www.uclalawreview.org/articles/archives/?view=55/6/1–7

Index

Page numbers in bold refer to figures and page numbers in italics to tables.